Security, Strategy, and the Quest for Bloodless War

Security, Strategy, and the Quest for Bloodless War

Robert Mandel

LYNNE
RIENNER
PUBLISHERS

BOULDER
LONDON

Published in the United States of America in 2004 by
Lynne Rienner Publishers, Inc.
1800 30th Street, Boulder, Colorado 80301
www.rienner.com

and in the United Kingdom by
Lynne Rienner Publishers, Inc.
3 Henrietta Street, Covent Garden, London WC2E 8LU

© 2004 by Lynne Rienner Publishers, Inc. All rights reserved

Library of Congress Cataloging-in-Publication Data
Mandel, Robert
  Security, strategy, and the quest for bloodless war / Robert Mandel.
    p.   cm.
  Includes bibliographical references and index.
  ISBN 1-58826-269-3 (hardcover : alk. paper)
  1. Casualty aversion (Military science). 2. United States—Military policy.
3. World politics—21st century. I. Title.
U163.M267 2004
355'.033573—dc22   2004003786

British Cataloguing in Publication Data
A Cataloguing in Publication record for this book
is available from the British Library.

Printed and bound in the United States of America

∞ The paper used in this publication meets the requirements
   of the American National Standard for Permanence of
   Paper for Printed Library Materials Z39.48-1992.

5 4 3 2 1

Contents

List of Figures vii
Preface ix

Introduction 1

1 The Quest for Bloodless War 7

2 Means of Pursuing Bloodless War 45

3 Precision-Guided Munitions 67

4 Nonlethal Weaponry 99

5 Information Warfare 125

6 What Can Go Wrong 155

7 Toward Effective Casualty Aversion 173

8 Security Policy Implications 187

Selected Bibliography of Works on Bloodless War 203
Index 205
About the Book 209

Figures

1.1	General Casualty Aversion Priorities	11
2.1	Strengths and Weaknesses of Casualty Aversion Means	64
3.1	Brute Force Versus Precision Weaponry	72
4.1	Classification of Nonlethal Weaponry	102
5.1	Information Disruption Versus Psychological Operations	129
6.1	Dangers of the Quest for Bloodless War	156
7.1	Conditional Utility of the Quest for Bloodless War	174
8.1	Security Policy Advice and the Quest for Bloodless War	188

Preface

This book is destined to stir up considerable controversy among readers. To question the trend toward relying on casualty-minimizing strategies as means of fighting wars and simultaneously to assert that fighting wars inevitably necessitates considerable human carnage are certain to raise hackles. Despite this risk, however, opening an intellectual debate on the subject seems crucial, especially in light of patterns in recent international armed confrontations.

Many observers assume that it is natural when waging war to attempt to minimize injuries and death, and further that political and military leaders have incorporated such casualty concerns prudently in making decisions about coercive actions overseas. These observers are likely to challenge the idea that there are important open questions surrounding the topic. In contrast, other observers assume that political and military leaders initiating wars care little about minimizing casualties, paying lip service to this humanitarian goal simply to appease the public. One outcome of this assumption is the contention that the only way to reduce or eliminate the human carnage of war is to terminate coercive action altogether. Even the scholarly literature on this topic has tended to shy away from its controversial dimensions: analysts typically either cheer the minimization of casualties during wartime as vital to successful military endeavors or make light of this quest for bloodless war as being unrealistic, unwise, or peripheral to mission accomplishment. Neglected are the rich and multisided insights possible from an in-depth examination of the appropriateness of casualty minimization as a priority for modern warfare. Thus it is precisely such an examination that I undertake in this book.

I certainly do not attempt to set forth general rules about the conditional sensibility of foreign military intervention or the initiation of warfare, nor do I assert that the avoidance of injuries or death should be the dominant determinant or the primary objective of warfare. Rather, I argue that casualty aversion—one element in the complex and multifaceted calculus of how to engage in foreign military intervention and warfare—has begun in recent decades to play an inappropriately large and sometimes misguided role in international confrontations. We need a more refined and differentiated understanding of the interplay of the different elements affecting casualty minimization, including (1) cost-benefit calculations about what is necessary for mission success, (2) signals to allies and adversaries about credible commitment, military advantage, and strategic vulnerability, and (3) ethical restraints on coercive activity overseas.

My underlying assumption is that the quest for bloodless war has raised absolutely central security and strategy issues, particularly in light of the changing post–Cold War international context, issues that cry out for fresh scrutiny. The unanswered questions surrounding the quest for bloodless war connect directly to profound questions about both the legitimacy and the effectiveness of the use of force in the current disorderly global system.

Clear dangers surround the seductive lure of thinking that one can wage and win modern wars without substantial costs in terms of human life (including both soldiers and civilians and both friends and foes). This book does reflect my fundamental assumption that winning wars always has and always will entail human sacrifice, a stark reality that must not be clouded by the availability of new technologies that appear to make obsolete the blood-and-guts confrontations of the past. It is critical not to see casualty aversion strategies as a panacea that can permanently eradicate the horrifying sting of violent conflict. There are circumstances where focusing on casualty avoidance is important, as well as occasions when such a focus interferes with mission accomplishment, and I attempt here to explore both sets of conditions without prejudice.

The impetus to write this book stemmed from a fascination with the seemingly contradictory U.S. desire in recent years to undertake frequent overseas military interventions and at the same time to minimize the loss of life on both sides of a violent conflict. My initial reaction to this aspiration to minimize the human costs of war was an intense curiosity about how it comes to pass, under what conditions this approach is viable, and its security implications in the current international system.

As with my previous books, my thoughts in this study reflect years of pondering. As part of my ongoing research, I have had conversations with numerous colleagues in both academic and policy settings who have significantly contributed to my thinking, and I especially wish to thank profusely for their help with this book the government officials toiling in defense and intelligence roles. I of course take full responsibility for any errors found here. This book is dedicated to all persons—civilians and combatants—who died as a result of warfare. It is impossible to write about the quest for bloodless war without feeling great compassion for the soldiers who died during combat, often at a very young age, and for the families who lost loved ones. Whether dove or hawk, no observer can fail to be touched and impressed by the heroism and valor of those who lose their lives on the battlefield.

Introduction

The desire to initiate war and emerge victorious while at the same time minimizing human harm is nothing new. Somehow achieving decisive victory without a drop of blood spilled is the ultimate political—if not military—fantasy. However, thanks primarily to advances in technology, in some people's minds we now appear to have the capacity to make that age-old dream a reality. The idea of actually being able to impose one's will through force around the globe without significant human costs is incredibly appealing—at least in part because of the underlying assumption that any opposition to such coercive action would emerge due to concern not that it was unwarranted or unjustified, but rather that it resulted in too many battle deaths.

Debate on the Issue

Many government officials, swept along by the optimism associated with modern high-tech offensive and defensive weapon systems, are very supportive of the quest for bloodless war and are attempting to allocate with concerted vigor substantial resources to promote strategies deemed likely to accelerate its pursuit. Similarly, many onlookers appear to utilize casualty minimization as their yardstick for success:

> In evaluating modern war, human rights groups, the press, and the public often look to the number of dead and wounded civilians as a meaningful metric. Civilian casualty figures sometimes are used to assess the morality, effectiveness, or legitimacy of military intervention. . . .

> [A]ssessing civilian casualties is an essential element of ensuring that Western publics assume informed responsibility for their government's use of military power.[1]

Few pause to ask the many fundamental questions about the ultimate feasibility or desirability of this quest for bloodless war. Furthermore, important concerns about whether one's rationale for military action overseas makes sense have the potential to take a distinct backseat to concerns about how many people are killed or injured in the fighting.

Although casualty aversion clearly has legions of staunch defenders, paradoxically its skeptics come from both the left and the right of the political spectrum. Left-wing critics argue that the quest for bloodless war is just a deceptive hypocritical sham, a pretext for military adventurism, degradation of public health, and prolonged conflict:

> There is no such thing as a "clean" war. The accuracy of "smart weapons" does not guarantee the safety of civilians, and it may even tempt field commanders to attempt to hit targets very near civilians. The tactic of "shock and awe" will cause catastrophic damage to essential urban infrastructure. "Collateral damage" is a euphemism for systematic disregard for the medium and long-term public health effects of destroying lifeline infrastructure, essential civilian services, and forcing the displacement of hundreds of thousands of civilians. It is hypocritical at the very least for U.S. military and political leaders to claim that their tactics are designed and implemented in a way that protects civilians.[2]

Some even see the quest for bloodless war as a warped ethnocentric justification for saving one's own people while indiscriminately slaughtering others. In contrast, right-wing critics (though more divided) contend that the quest for bloodless war replaces battlefield courage with cowardice and prevents the military from undertaking the concerted use of overwhelming force in foreign confrontations necessary to achieve decisive victory and overall national security.

Broad Scope of This Study

Given the scarcity of broad integrative treatments of the quest for bloodless war, this study provides a comprehensive analysis—incorporating both moral and pragmatic perspectives—of many tantalizing conceptual conundrums surrounding this topic. The book incorporates an exploration of the origins and history of the notion of bloodless war, including debates about its definition, morality, and expedient value; the various means of pursuing casualty aversion; developments surrounding three central instruments of casualty aversion—precision-guided munitions, nonlethal weaponry, and information warfare—including pertinent case evidence from the 1991 Gulf War, the 1999 conflict in Kosovo, the 2001–2002 war in Afghanistan, and 2003 war in Iraq; the dangers of overreliance on casualty aversion; the conditional utility of the quest for bloodless war; and the important security policy implications. Throughout, I attempt to maintain balance in the treatment of divergent perspectives.

The focus of the book is on a largely deductive conceptual exploration of casualty aversion during warfare, highlighting controversies and trade-offs surrounding the issue and looking especially at its long-term political and military implications. The most important objective of the analysis is to isolate the operational utility of casualty aversion for security policy.

The questions raised encompass such intertwined national-security topics as the nature of modern warfare and battlefield strategies, civil-military relations, the value of rapid technological developments, the clarity of international communication, the prerequisites for international legitimacy, the trade-off in the security realm between morality and effectiveness, the role of democracy in restraining coercive activity, and the emergence of global security norms. More specifically, attempting to minimize casualties during warfare can signal that battlefield strategies entailing major sacrifices in human life may be less tolerable than in the past; that the ethos of the soldier may be changing and the gap between civilian and military values decreasing; that military power may increasingly rest (and be perceived as resting) on possession of key advanced technologies rather than on trained soldiers; that willingness to suffer casualties may no longer be as crucial to communicating credibility; that the global legitimacy of one's actions may rest more on the means one uses to fight a war than on the ends one hopes to achieve; that adhering to a strict moral code in warfare may entail greater ineffectiveness against the types of foes faced today than those faced in the past; that the spread of democracy around the globe may be leading to an unprecedented acceleration and harmonization of concerns about casualty aversion; and that, consequently, despite cultural differences, advanced industrial societies' security norms may be converging. Thus an examination of the quest for bloodless war serves as a springboard for taking a fresh look at a wide range of vital yet contentious national security issues.

Although this book makes every attempt to utilize the broadest conceptual lens in examining the quest for bloodless war, three specific emphases characterize its exploration. First, despite inclusion of comparative reference points wherever possible, the analysis tilts toward the United States because of its recent dominant global security position and articulated emphasis on minimizing wartime casualties. Second, despite extensive references to earlier historical periods, the focus is on post–Cold War application of casualty aversion strategies (including changes after September 11, 2001), because the underlying desire is to gain insights from the time period of greatest relevance as they relate to coping with emerging security problems. Third, this book concentrates on international warfare and major foreign military interventions, rather than internal conflicts or international peacetime activities, because international violence exhibits the greatest complexities, contradictions, and costs in attempts to minimize loss of life.

Timeliness of This Investigation

Looking at the quest for bloodless war now seems particularly important because the wave of unbridled advocacy currently supporting this aspiration appears to be receiving concrete reinforcement from recent military successes that demonstrate the feasibility and effectiveness of combining casualty aversion with decisive military victory. Looking back historically, "the idea of bloodless wars on the battlefield of the future has been the subject of much debate among war strategists for a long time";[3] yet today, more than ever before, to many observers "the benefits of bloodless battles for soldiers and law enforcement are obvious."[4] That many analysts currently do not even see the controversy surrounding an emphasis on casualty aversion is reason enough to undertake this study.

The post–Cold War environment has fostered a setting where major powers like the United States do not always have a clear formula about when to intervene militarily in troublesome predicaments around the world and, if they do intervene militarily, how to judge the success or failure of a particular war or intervention. The prioritization of what is really important to U.S. security has in some ways become so muddled that the United States seems reluctant to identify explicitly any crisis in international relations as truly worthy of major sacrifice on its part. Reinforced by its reduced post–Cold War sense of obligation to address global problems, the United States appears—despite forceful action during the recent war on terrorism and the war against Iraq—to retreat to a position where it attempts to maintain global dominance and keep "truants" in line while minimizing its risk or commitment to engage in unabashedly protracted bloody confrontations.

This ambiguous and confusing context opens the door to casualty minimization concerns, often playing an increasingly unwarranted role as an inadequate proxy or substitute for both a sound rationale for military commitment and a sound measure for successful military action. The justification for many recent foreign security policy initiatives appears to give undue weight to their minimization of casualties compared to the obviously more important issue of their accomplishment of objectives to change the situation at hand. To many observers, quite inappropriately, wars or foreign interventions that are warranted but that result in sizable casualties are often deemed failures, while those that are not warranted but that result in few deaths are frequently deemed successes. Clearly there are major problems here, and there is a need to get at the root of the justification for casualty aversion and the thrust of international military action.

Projected Audience

This work is designed to speak to both international relations scholars and national security policymakers about the opportunities and limits surrounding the conduct of modern warfare. The passion behind this volume arose in part from a desire to try to help those who are responsible for grappling with a complex set of emerging threats. Bridging the gap between academic and government security studies is absolutely crucial to increase sensitivity on both sides to the different perspectives involved and to allow each group to benefit directly from the findings of the other. Thus this study makes every effort to avoid scholarly and policymaking jargon, unexplained acronyms, or the implicit assumptions of any supposedly prevailing school of thought.

Notes

1. Project on the Means of Intervention, "Understanding Collateral Damage" (Washington, DC: Harvard University, John F. Kennedy School of Government, Carr Center for Human Rights Policy, June 4–5, 2002), p. 2.

2. Ben Wisner, "Notes on the Ideas of 'Clean War' and 'Collateral Damage,'" unpublished paper, March 17, 2003, p. 2.

3. BBC News, "The Weapons of Bloodless War" (May 13, 2003), available online at http://news.bbc.co.uk/1/hi/technology/3021873.stm.

4. Douglas Pasternak, "Wonder Weapons," *U.S. News & World Report* 123 (July 7, 1997): 38.

1
The Quest for Bloodless War

International warfare has changed considerably in principle and practice over the years. Yet throughout the history of armed conflict, one constant has been the concern about human casualties, albeit with considerable fluctuation in emphasis, interpretation, and underlying justification. From whence does this concern about casualties emerge? What kinds of motives have surrounded its persistent reemergence? How does the public feel about the issue? What is the historical pattern of its application? Most central, why is it that there is a renewed aspiration to achieve bloodless war?

A starting point is the recognition of a fundamental paradox evident from the history of warfare surrounding the relationship between lethality and casualties: "It is clear that as weapons have become *more* deadly, the results of exposure to standard weapons concentrations for standard periods of time have become *less* deadly, since the rate at which weapons inflict casualties on the battlefield has been declining."[1] The standard explanation of this pattern is simply that "the increased dispersion of soldiers has increased more rapidly than weapons lethality per person";[2] yet this paradox opens the door to an endless stream of alternative explicit and implicit rationales, including a growing revulsion at massive wartime loss of life. Nonetheless, this paradox serves to highlight how bizarre it is that—at the same time—civilized society tries to maximize its capacity both to destroy and to save lives during battle.

This chapter takes a comprehensive, detailed, and multifaceted look at the nature and origins of the quest for bloodless war. After considering definitional ambiguities, the chapter examines the most fundamental roots that emerge from the contrasting motivations of idealistic humanitarian values (including moral principles, democratic beliefs, and international law) and expedient practical needs (including external support, media coverage, and technology advancements). Lurking beneath these elements is the perceived casualty sensitivity of the public. From an examination of the actual historical record of concern about loss of life, it is clear that its salience has intensified since the end of World War II. The United States has placed a special emphasis on casualty aversion, especially since the end of the Vietnam War, but is by no means alone among advanced industrialized countries in its concern about this issue; yet throughout much of the third world, a diametrically opposed pattern is present. Finally, considerable controversy surrounds the impact on casualty aversion of the September 11, 2001, terrorist attacks on the World Trade Center and the Pentagon.

Definitional Issues

The notion of casualty aversion—also commonly called casualty avoidance, dread, minimization, sensitivity, or phobia—operationally means that during warfare one has a low tolerance for losing many lives or suffering many injuries. Attempts abound to define precisely this elusive concept (whose opposite would be called casualty tolerance or acceptance), such as "an unwillingness by the political or military leadership to place the American military in a position of danger, even to the exclusion of accomplishing policy aims."[3] Some observers distinguish casualties from the broader category of "collateral damage," which they define as involving both "nonmilitary structural damage and human casualties that occur in the course of striking valid, approved military targets."[4] Other analysts attempt to create a scale of intensity regarding this issue, placing, for example, casualty phobia—demanding virtually no loss of life or a "zero-casualty" mind-set—rather than casualty aversion at the more extreme end. Still others strive to pin down a formula determining a precise threshold for the number of acceptable casualties in different types of military confrontations. Overall, however, a universally accepted precise meaning for casualty aversion has remained elusive.

In many ways, the quest for bloodless war represents an aspiration, embodying a set of sometimes unspoken or confusingly stated motivations, intentions, and values, rather than a pattern of unambiguous empirically observable behavior. One could be making a sincere effort to minimize casualties during war, and yet inadvertently cause them; or alternatively one could be an indiscriminate killer during war, and yet not end up hurting anyone. Furthermore, it is very difficult to obtain reliable data on deaths and injuries among all relevant parties during wartime. In recent conflicts,

> civilian casualty numbers often have been inflated by the opposing force, ignored by the intervening military power, and imperfectly reported by the media and human rights organizations." ... [T]he U.S. military rarely provides collateral damage figures during conflicts, and often is reluctant to comment on allegations of collateral damage for lack of comprehensive and accurate information.[5]

In many cases, public awareness of casualties—particularly among adversaries—is low, well exhibited during the 2003 war against Iraq, for which the U.S. government did not release any reliable official figures about Iraqi civilian or military deaths resulting from the combat. Without solid information, it is tough to refine what kinds of operational indicators might be best for what purposes. Tightly connected to the paucity of reliable casualty data during warfare is the difficulty of determining whether focusing on casualty aversion leads to expected long-term benefits.

Definitional ambiguity characterizes even the most basic components of the quest for bloodless war. Disagreement surrounds the issue of what constitutes a combatant, a particularly muddied question when it comes to nonstate foes such as terrorists: Are combatants restricted to official enemy soldiers, wearing the distinctive garb of military forces, or does the notion of combatants include anyone who picks up a weapon to fight you or forcibly tries to resist your advances? Interestingly, a key distinction separates two types of governmentally sanctioned combatants: soldiers fighting in international wars generally have much greater freedom to engage in courses of action where innocent civilians will knowingly be harmed than do domestic police. Furthermore, even the concept of casualties itself is murky—to some, for example, the term *casualties* includes "not only the killed and wounded but also losses from disease, desertion, accidents, and troops taken prisoner or missing in action."[6] Such ambiguities can indirectly thwart the formulation of sound casualty aversion policies; for instance, since in past eras "disease has been the biggest killer of soldiers,"[7] far more deadly than enemy bullets, does this mean the quest for bloodless war ought to concentrate on expanding the availability of medical assistance?

Beyond human incapacitation and death, many onlookers are equally if not more concerned with other kinds of devastation resulting from warfare. Among the many alternative emphases are damage to civilian infrastructure, disruption of the economy and livelihoods, and even degradation to the ecology of the region.[8] For some, in other words, the bombing in enemy territory of unoccupied schools, hospitals, museums, or vital non-defense-related computer facilities or manufacturing plants, or alternatively the destruction of a precious fragile ecosystem, can elicit more shock and outrage than the killing of people.

Thus controversy surrounds the question of what type of damage is or should be the subject of greatest concern. For onlookers who focus on human carnage, all deaths or injuries witnessed on the battlefield do not count equally. While there is usually concern about avoiding the death of significant numbers of one's own military troops (commonly termed "force protection") or of fellow private citizens, minimizing enemy troop or civilian casualties may not be a priority. Within the United States, for example, controversy surrounds public casualty aversion priorities. Some argue that "although many Americans are extremely sensitive to American casualties, they seem to be remarkably insensitive to the deaths of foreigners including essentially uninvolved— that is, innocent—civilians";[9] in contrast, others contend that "significant segments of the U.S. population demonstrate major concern for minimizing risk to adversary civilians."[10] Even with the long U.S. tradition of concern about loss of life, with military institutions educating their students to be sensitive to enemy casualties, large fluctuations in concern across foreign military confrontations are evident. In situations where the country has been a victim of an attack perceived as vicious and unprovoked, such as the 2001 terrorist attacks on the World Trade Center and the Pentagon, it is certainly easy to dehumanize and indeed demonize foreigners and engage the desire to launch vengeful and bloody retaliation that ignores any possible suffering among the perpetrators and their supporters. In contrast, in situations where no such attack on one's homeland has occurred, such as the U.S.-led effort to compel Serbian leader Slobodan Milosevic to cease ethnic cleansing in Kosovo in 1999, the concern about minimizing enemy casualties seems much higher.

Nonetheless, across countries, time periods, and violent conflicts, it appears possible to specify a general prioritization of concerns relating to the quest for bloodless war. For both leaders and the mass public, concern about casualty aversion appears to be greatest for one's own civilians, followed by one's own soldiers, then foreign civilians, and—finally—foreign soldiers. Figure 1.1 displays this important hierarchy. It also seems reasonable to posit that the worries about human destruction still take precedence over nonhuman destruction (structures, information systems, and the physical environment) during warfare.

Idealistic Humanitarian Motives

Looking across the broad sweep of human history, protection of human life has not traditionally been the highest national security goal of the sovereign state.[11] Protection of territory and protection of regime have usually taken precedence over such humanitarian concerns, with many rulers explicitly deeming the lives of their citizens—whether soldiers or civilians—expendable. However, the advent of more representative forms of government in which leaders have become more accountable to the people, combined with the refinement of norms and rules guiding behavior during warfare, has gradually altered this long-standing priority system.

Moral Principles

The rationale for the quest for bloodless war often reflects a high-minded ethical justification for protecting civilians during wartime. Principles deriving from moral philosophy can be powerful grounds for

Figure 1.1 General Casualty Aversion Priorities

	Civilians	Soldiers
One's own country	1	2
One's major enemy	3	4

Note: 1 = highest security priority, 4 = lowest security priority in the quest for bloodless war.

restraining the behavior of soldiers and for having casualty aversion play a major role in policies governing military confrontations overseas. Even though these principles are largely intangible and devoid of enforcement, and even though many states disagree on both the identity and the meaning of these principles, they can still contribute to the perceived legitimacy of coercive security actions abroad.

The criteria of just war theory surrounding the employment of force in many ways form the basis of this moral perspective. According to these criteria, if a state goes to war, then it needs to concern itself in its conduct during the conflict with proportionality of means and noncombatant protection: during combat the state must employ no greater use of force than is absolutely necessary and do as little damage as possible to noncombatants. The first principle serves to minimize casualties among allied and enemy soldiers by prohibiting gratuitous or unnecessary harm to fighting forces, while the second principle minimizes casualties among allied and enemy civilians, with the past hundred years particularly demonstrating a growing international consensus that seeks "to protect from the ravages of war whole classes of people not directly involved in the prosecution of war."[12] The most important formal effort to restrain the destructiveness of war, on moral grounds, "has taken place in the modern period" because of "the realization of war's increasingly devastating capabilities and the use of armed power during war to attack civilian noncombatants."[13] The basis in just war theory for the protection of human life—of soldier and civilian and friend and foe alike—is that "the rules of 'fighting well' are simply a series of recognitions of men and women who have moral standing independent of and resistant to the exigencies of war," with a legitimate act of war being "one that does not violate the rights of the people against whom it is directed."[14] The greatest moral sanctions are against intentional rather than accidental occurrences of unnecessary harm during warfare.

Invoking concerns of morality seems especially likely on the part of today's liberal regimes:

> In the wake of the September 11th terrorist attacks in the United States which killed about 3,000 innocent civilians, President George W. Bush argued that only barbarians target civilians, whereas civilized people—presumably those residing in Western liberal democracies—respect human rights and avoid hurting noncombatants. . . . Theoretically, the liberal beliefs and democratic institutions of such states suggest that they should be careful not to kill many civilians in war. Liberal norms, for example, prohibit the harming of innocent individuals,

even enemy civilians in wartime. Democratic institutions, on the other hand, force leaders to be mindful of public opinion in making foreign policy choices. Just as fighting a costly war—or even worse, a losing one—is a policy likely to result in a leader's repudiation at the ballot box, killing large numbers of civilians in combat operations is liable to provoke public censure, possibly leading to loss of elected office for the officials responsible. Finally, liberal democracies are presumably the type of regime most sensitive to international ethical norms prohibiting intentional or disproportionate harm to noncombatants, since democracies themselves abide by similar norms domestically and advocate and propagate them internationally.[15]

The open liberal international system provides a set of clear aspirations to minimize bloodshed.

The United States in particular has repeatedly made reference to the moral underpinnings of the quest for bloodless war. The U.S. government has traditionally believed that "respect for human life and consequent casualty consciousness are fundamental to what makes a nation civilized"; that, given its position as a superpower, "the sanctity of human life in time of war or peace certainly ought to be" the country's "foremost example to the world of how to act"; and that "it is therefore a moral issue that leadership must be casualty conscious in prosecuting war, for technology has turned acceptable wastage rates of the past into today's indictments of horrifying military incompetence."[16] The country thus in many ways makes explicit its desire to exemplify and expound to the world a set of civilized and humane norms during warfare.

Democratic Beliefs

The principles of democracy inherently embody respect for the sanctity of human life, and as this form of government has spread around the world, so has this human rights value. Although few if any states may live up to the lofty ideals of pure democracy, and even though regimes recognized as being largely democratic may in reality be just as likely as nondemocracies to kill noncombatants purposefully as part of wartime political-military strategies,[17] there is little doubt that this form of government associates at some level with concern about human suffering. To win wars, democracies need to maintain popular support, and to maintain popular support, democratic governments must to some extent be sensitive to human casualties, and so casualty aversion seems to be a virtually inevitable consequence because people would be upset if many of their fellow citizens were being slaughtered in battle.[18]

Furthermore, the presence of democracy induces a special kind of restrained calculus in committing the military to action overseas:

> In a democracy, policymakers contemplating the deployment of troops into a situation that is, or might become, hostile, should sensibly evaluate three considerations insofar as they desire support for the action from the public. First, they must consider the value the public places on the venture, and they may try to use whatever persuasive skills they possess to enhance this value—that is, to sell the project to the public. Second, they must consider the likely costs of the venture, particularly in American battle deaths. And third, they must evaluate the potential of the political opposition to exploit the situation should battle deaths surpass those considered tolerable by the public.[19]

Indeed, modern democracies may fail to achieve their goals in warfare because they are unable "to find a domestically acceptable trade-off between brutality and sacrifice"—unable to forge a winning balance between expedient and moral tolerance for the costs of war—with educated citizens abhorring the brutality of limited overseas coercive engagements but at the same time refusing to sustain the level of casualties resulting from fighting in other ways.[20] Thus the restrained fighting calculus induced by democracy, increasing the odds of internal legitimacy, may not always lead to the most efficient military operation.

International Law

Moral principles and democratic beliefs have contributed to the development of formal international law protecting the rights of civilians during war. The "laws of war" amount to "a body of rules that the world's nations have collectively crafted over 137 years of conventions in Geneva and The Hague—but enforced with only sporadic success":

> The idea that certain fixed laws should apply even amid the violence and anarchy of war isn't new. The saying may have it that all's fair in war, but restrictions on battlefield conduct have always been recognized. The Hebrew Bible forbade soldiers from, among other things, destroying fruit-bearing trees in hostile lands, and chivalric codes existed in the Middle Ages.[21]

Thus international law has provided a concrete starting point for the concerns surrounding casualty aversion and human suffering during war, with a particular emphasis on protecting noncombatants from adverse effects.

Reference to the Geneva Conventions in particular has surrounded the concern about protecting civilian lives during warfare, even though "the record shows that the ratio of civilians to military personnel killed in armed conflicts has, in fact, *increased* since the [Geneva] Conventions of 1949."[22] One indication of the continued relevance of the Geneva Conventions, despite the frequent violation of their precepts by many states, is Iraq's attempt to invoke this body of law during 2003 to try to restrain U.S. military actions:

> As Saddam Hussein's operatives and the Iraqi military take asymmetrical warfare to the extreme, Baghdad is using U.S. adherence to the rules of the Geneva Convention in Iraq almost as a weapon in its arsenal. Iraqi fighters are trying to force unpalatable decisions on U.S. commanders far less willing to kill or threaten civilians than the Iraqi military itself. . . . Iraqi soldiers have waved white flags to lure an enemy into an ambush, something that's clearly proscribed under the Geneva Convention, which also bans firing from behind human shields of women and children—another Iraqi tactic. Likewise, storing munitions in hospitals and setting up mortar and machine-gun positions on the roofs of schools and mosques isn't considered acceptable by any conventional measure of wartime engagement.[23]

In other words, countries battling the United States and Western military coalitions may try to exploit these advanced industrial societies' respect for the international rules of war—"their goal is to gain political leverage by portraying U.S. forces as insensitive to LOAC [law of armed conflict] and human rights."[24] In addition, these Western states may occasionally choose to impose these legal restrictions on themselves: during the 1999 Kosovo conflict, "US officials are reported to have decided against deploying their electronic arsenal because of fears that the impact on civilian life would have led to charges of war crimes under the Geneva Convention."[25] Thus international law can affect the number of casualties by restraining both the ways states fight wars and the ways target states attempt to manipulate attackers perceived as abiding by the rules.

Qualifications

There is, of course, an inherent tension between the global spread of morality, democracy, and law, and the global condition of anarchy—without overarching common security values—that characterizes today's world. In many senses, the quest for bloodless war is caught between these two: the values associated with morality, democracy, and law induce a concern about civilized behavior and protection of human life, reinforced by high levels of globalization and interdependence; but the anarchic structure promotes a kind of "might-makes-right" exercise of power that concerns itself with nothing more than whether someone else is strong enough to stop you or punish you for what you are doing. This second, more chaotic perspective is especially exacerbated by the proliferation of disruptive nonstate forces such as international terrorists, the rapid diffusion of weapons technology across porous national borders, and the acceleration of violent low-intensity conflicts across the globe.[26] So frequently in today's international confrontations, only one of the sides has a real concern about casualty aversion, enhancing the perceived asymmetry of will.

Furthermore, minimizing military casualties through force protection—a key thrust of modern casualty aversion—does not always represent fulfillment of the moral, democratic, or legal ideals either for a civilized society or for the military establishment itself:

> Despite the accuracy of the air attacks, too many civilians were killed while allied combatants avoided risk. . . . This turns a principle of a just war on its head—specifically, the obligation to protect the innocent at the expense of the warrior. Another troubling and similar aspect of the so-called "immaculate" air campaign is the ability to drive an enemy to his knees without shedding a drop of the bomber's blood. Normally, the litmus test of going to war was the willingness to suffer casualties in pursuit of its objective.[27]

This issue highlights the potential trade-off between protecting the lives of civilians and protecting the lives of combatants.

Expedient Practical Motives

Aside from these largely humanitarian concerns, from the point of view of military expediency a few practical motives are associated with minimizing loss of life. The pursuit of casualty aversion often can reap support for one's cause from foreign governments, international and transnational organizations, and the domestic public; generate glowing media coverage amplifying the positive spin-offs during and after warfare; and take advantage of the latest technologies to give one's state a military advantage. In each of these cases the avoidance of negative costs is at least as important as the achievement of positive benefits. Substantial political payoffs may emanate in today's world from the quest for bloodless war, even to leadership that may not genuinely care about the welfare of its citizenry.

External Support

Undertaking compassionate and discriminating conflict abroad can create a positive glow in several ways. The United Nations and transnational human rights and peace organizations would be pleased that the rights of enemy civilians and soldiers are being respected, as would observer states that either sympathize with an enemy's cause or are not necessarily convinced of the wisdom of a military action. For example, in the 2003 war against Iraq, "a large part of the reason why the U.S. forces wanted to limit civilian casualties was that they were fighting a battle for hearts and minds—primarily in Iraq but also in the rest of the world."[28] In the war on terrorism in Afghanistan in 2001–2002, casualty aversion was important because "mass casualties could inflame anti-U.S. sentiment around the world, particularly in Muslim nations, and convince many people that America is engaged in a war on Islam; they could also make it difficult for Arab or Islamic governments to cooperate in the broader campaign against terrorism, and even put friendly governments at risk of being toppled if popular anger erupts over their logistical or political support for the U.S."[29] Within one's own country, minimizing foreign casualties can please domestic pacifists and groups who are ethnically, religiously, or ideologically linked to prospective victims in other countries, a particularly important benefit given the rising mobility of people and creation of ethnic diasporas. For example, due to the presence of a large Muslim population within its borders, France has a strong motivation to be very restrained in killing North Africans during conflict in Algeria.

Moreover, striving to minimize casualties among one's adversaries can result in significant postwar benefits. In the 2003 war against Iraq, for instance, collateral damage was minimized to preserve Iraq's infrastructure, so that after the conflict "the country's civilian population and what's left of its armed forces and political structure" would be "willing and able to organize and rebuild under temporary occupation by foreign (mostly US) forces."[30] Casualty aversion can thus dramatically reduce reparation costs in terms of both physical reconstruction of buildings and political resocialization of the population.

Media Coverage

The greatest notoriety for spurring the quest for bloodless war undoubtedly goes to the media, even though much of the evidence for the impact of this element is speculative and heavily intertwined with other causes. The most commonly attacked media culprit is television, due to its special ability to dramatize the costs of war.[31] Part of the reason Vietnam casualties are still remembered today is because of their unprecedented visibility during the conflict. Vietnam was the first war watched on television and "was a forerunner of what would later be called 'the CNN effect.'"[32] During the Vietnam War, many analysts believed that the bloody film clips on television were responsible for triggering casualty sensitivity among the American people.

During more recent wars, with even more extensive, graphic, and intimate coverage by the media, this impact intensified. According to a senior military official, "We have entered the 'Visual Era'; our rules of engagement are new—dominated by the risk that parents will see their sons and daughter killed in real-time on TV . . . we can't afford casualties."[33] The information revolution, when combined with the proliferation of democratic regimes, has enlarged the media impact:

> It spawned high-tech global news organizations that rapidly deliver information—to include graphic images of war—to publics everywhere. This is particularly important when considered in conjunction with another attribute of the information age: the spread of democracy. Shaped by raw news footage, public perceptions of how conflicts are being fought significantly affect military interventions. . . . In modern popular democracies, even a limited armed conflict requires a substantial base of public support. That support can erode or even reverse itself rapidly, no matter how worthy the political objective, if people *believe* that the war is being conducted in an unfair, inhumane, or iniquitous way.
>
> The velocity of today's communications' capabilities presents real challenges to democracies as well as those governments that if not democratic in a Western sense, nevertheless depend upon support from constituencies having access to globalized information sources. When television airs unfiltered, near real-time film of what appears to be LOAC [law of armed conflict] violations, complications result.[34]

Thus the combination of global news transparency and democratic values can reinforce a casualty aversion emphasis.

Mass media can sometimes foster during warfare not only government use of casualty aversion strategies but also the receipt of praise for such restraint. Having independent television, newspaper, or radio outlets demonstrate the civility of one's wartime behavior can be far more credible than any statements by the governments directly involved. Such media praise helps to avoid situations where, after a war, "you lose the peace" due to broadcasts highlighting civilian casualties and collateral damage.[35]

Technological Advancements

Developments in weapons technology have underscored the importance of the quest for bloodless war to security policy. In the U.S. tradition especially, "technology is our first answer to the lethal hazards of waging war."[36] The globally accelerating pace of innovation in military technology, possessing increasing destructive potential, has highlighted the casualty aversion issue: "The unprecedented carnage of 19th-century warfare led Western nations to try to mitigate the most unnecessary forms of battlefield suffering" because "innovations in weaponry and the advent of 'total war,' which exacted unconditional surrender from the defeated party, had made combat deadlier than ever before."[37] The fluid offensive-defensive arms development cycle has spurred continuous innovation in armaments, including the capacity to discriminate better among targets. In the twentieth century, "there has been continuous, rapid growth in the reach, lethality, speed, and information-gathering potential of armies,"[38] thus making it more possible to wage war without spilling a lot of blood. "Technological determinism" may be involved here, in which the presumed aversion to casualties by Americans, for example, occurs "because they *can* avoid them, because America's overwhelming technological advantage allows it to dominate the enemy and fight on its terms."[39] Indeed, technological advancements serve to support the casualty aversion value both because modern high-tech war fosters devastating carnage and because "precision-guided weapons and stealth make casualty limited warfare possible."[40] The military may thus be increasingly tempted to rely more on new technologies—rather than innovative strategies and tactics—for success on the battlefield. However, it is still the case that "high casualty figures reflect not only the lethality of the weapon inflicting them but also the tactics employed on both sides,"[41] particularly the degree of ruthlessness among the human combatants.

Although the costs of casualty-minimizing weapons technologies have plummeted in recent years, they are still often more expensive than their more indiscriminate counterparts. For a major power like the United States, facing escalating global military commitments can trigger the kind of calamitous budget deficit that makes it difficult to have the kind of preferred casualty-minimizing munitions in every place for every contingency. The inevitably resulting discussions about "security on the cheap" can pose a serious practical limitation, even within wealthy states, on universal reliance on technologies useful for casualty aversion.

Even though casualty minimization can be achieved through military initiatives launched from land, sea, and air, most observers link the increasing reliance on air power in recent wars to reduction in battle deaths. For example, an adviser to senior Pentagon officials said that, during the 2001–2002 war against terrorism in Afghanistan, "concerns about high U.S. casualties led the Bush administration to craft a strategy that relied on air power and small numbers of commandos, as opposed to tens of thousands of American ground troops."[42] Although in certain circumstances sea or land initiatives possess real advantages for casualty aversion, proponents of air power have been among the quickest to jump on the bandwagon of the quest for bloodless war.

The promise of high technology is driven by the desire to (1) minimize collateral damage so as to assuage perceived public concerns and increase legitimacy; (2) achieve force protection through increased reliance on machines that can be effective from safer distances; and (3) attain a presumed "quantum leap" in cost effectiveness.[43] However, this last premise is not without controversy:

> All these hopes are questionable, however. High-tech applications tend to be over-complex and, thus, susceptible to "Murphy's Law." High rates of mission capability are maintained only by Herculean maintenance efforts. Moreover, these systems, like all others, are vulnerable to counter-measures—but their high cost and long development cycles impede any quick adaptation to such counter-measures.[44]

Thus using advanced technology in bloodless war may involve some inescapable trade-offs.

An important underlying causal question concerns whether the development of technologies useful for casualty aversion triggers the advocacy for bloodless war, or alternatively, whether a preexisting desire to reduce carnage lead to the development of these technologies. Reflecting on the historical development of weapons technologies and the security policy discussions revolving around the wartime casualty issue, it appears that concern about casualty aversion occurred largely prior to (and was indeed a stimulus of) the development of these technologies. In other words, technologies with the capacity to minimize wartime loss of life would probably not be used in that way were it not for preexisting concern about casualties.

Qualifications

Orienting one's foreign security policy through casualty aversion in such a way as to maximize support from foreign countries and global organizations can of course be an unwise proposition, placing international political popularity and acceptance above effectiveness in accomplishing military missions. External credibility could suffer here as well. There is also always the risk that during warfare some onlookers will malign one's motivations for minimizing loss of life, or utilize an impossibly low threshold for determining whether the number of casualties actually generated is acceptable.

Media impact can also certainly work at cross-purposes during warfare: the media-induced "body-bag effect" induces "the fear of policy-makers that, once casualties are taken, public support for an intervention will rapidly wane"; yet this body-bag effect may sometimes operate "in opposition to the CNN effect, with one pushing policy-makers to intervene during a humanitarian crisis and the other urging caution against involvement in one."[45] The media's greatest impact on shaping military strategy regarding casualty aversion occurs if most of the major public exposure pushes a government in the same direction. Furthermore, despite claims to the contrary, considerable controversy exists about whether "the public responds viscerally to gruesome images on television and whipsaws policymakers with wild swings in mood."[46]

Technological advancements in war-fighting systems similarly constitute a two-edged sword both for minimizing loss of life and for achieving military objectives. More sophisticated weapons systems can discriminate better among targets, but at the same time often pack such a punch that collateral damage—including human casualties—can be more likely. Similarly, technological developments can improve one's reach to hit key control nodes deep into enemy territory, but at the same time, thanks to possible technological diffusion, make one's own command-and-control systems—increasingly electronic and tightly networked—more vulnerable to devastating attack.

Moreover, the practical expediency motives afforded by technological advancement may undercut the moral basis for pursuing casualty aversion. A look at the historical development of armaments indicates that "it is the inventive genius of man which has obliterated his sense of moral values," with weapons development in particular being the key to the dissolution of morality through endowing humans with the power to destroy on a large scale.[47] In other words, as increasingly sophisticated armaments increase the technological capacity to achieve military objectives without killing a lot of people, at the same time they may inadvertently decrease the moral desire to do so. Serious questions emerge about how technology affects the humanity of warfare:

> Will technology be used to make war more humane? If we read the question to mean, "Will technology result in wars that have fewer casualties and less collateral damage?" the answer is yes, almost certainly. If we read the question to mean, "Will technology result in less frequent wars fought for more 'noble' ends?" the answer is no, with an even higher degree of certainty.[48]

Advanced weapons technology distances initiators of violence from witnessing the direct suffering of targets, lessening the probability of moral inhibitions entering the picture.

Public Opinion and Casualty Sensitivity

Underlying both the humanitarian and expedient motives discussed is the issue of public attitudes about wartime casualties. An initiator's own civilian citizenry generally feels more patriotic knowing that its safety is a priority, and an emphasis on force protection can usually keep one's soldiers happy, loyal, and efficient in undertaking vital defense tasks. Yet tremendous controversy and misperception surround the question of just how casualty-sensitive the public really is. Many observers assume that body bags trigger the growth of domestic antiwar sentiment: "Political and military leaders justify this preference for sanitized, high-tech warfare by arguing that the American people simply will not tolerate casualties anymore, at least not in a 'small war,' that as soon as the body bags start coming home support for a mission will evaporate. This is said to be a fact of life in the television age."[49] The claim that "policy makers and senior military leaders believe that the American public is casualty-averse and will not tolerate deaths except when vital interests are at stake" receives support from isolated pieces of evidence: for example, a study comparing public support for the Korean and Vietnam

Wars reveals that the public was indeed sensitive to casualties and withdrew support based on cumulative number of casualties.⁵⁰ In contrast to past reactions such as during World War II, the presumption now links the sight of carnage to war-ending revulsion.

Yet recent studies have led to a startling contradiction of these prevailing beliefs, implying that the government may be misreading or misrepresenting public sentiment about wartime casualties. For example, a comprehensive Triangle Institute for Strategic Studies survey discovered that the American public is far more casualty-tolerant than are policymakers or senior military officers. More specifically, the survey rejected as a myth the common belief—"widely accepted by policy makers, civilian elites, and military officers"—that the American public is especially casualty-phobic and demands "a casualty-free victory as the price of supporting any military intervention abroad."⁵¹ Such public tolerance revolves around perceptions that (1) casualties are necessary to accomplish a military mission, (2) a state's leadership is strongly pushing for a military mission, (3) a military mission promises to be successful, and (4) a military mission is vitally important to national security.⁵² Accordingly, "the public does not demand bloodless interventions as the starting point for securing national interests and exercising world leadership,"⁵³ and public attitudes "reveal no irrational or knee-jerk reactions based on a putative unwillingness to tolerate casualties; they reveal no immature demand for a casualty-free security policy."⁵⁴

Indeed, some argue that, when popular concern arises, the public may respond to casualties with a renewed determination to persevere or even a desire to escalate the fighting:

> Moreover, polling data suggests that if the elites in government and the media are united in favor of a mission—as they were, for instance, during the early days of Somalia—the public is willing to go along, even if the mission does not conform to the dictates of "national security," narrowly defined. . . . In fact, suffering casualties can actually increase American public support for a military operation, so as to ensure that their soldiers did not die "in vain." Thus when leaders choose not to enter a war or choose to fight with the most antiseptic means possible, they should not use "public opinion" as an alibi for their actions.⁵⁵

Furthermore, "polls show little evidence that the majority of Americans will respond to fatalities by wanting to withdraw U.S. troops immediately and, if anything, are more likely to want to respond assertively" with continued military force.⁵⁶

Combat soldiers themselves may not press for casualty minimization:

> Those whose lives are on the line do not ask for a no-casualties policy. In one survey of 12,500 service members conducted by the Center for Strategic and International Studies [CSIS], 86 percent agreed with this statement: "if necessary to accomplish a combat/lifesaving mission, I am prepared to put my own life on the line." The CSIS study also found that "excessive aversion to casualties has . . . led to some confusion in the military, where self-sacrifice and the willingness to accept casualties in military operations has always been a key part of the ethos." Nor are most of the rank and file intrinsically opposed to "humanitarian" missions; many report that they like helping people and . . . express frustration with force protection policies that keep them from doing more.[57]

These members of the armed forces expect and even desire to risk their lives in combat action. However, there appears to be some difference between the attitudes of junior and senior officers: "Senior officers with memories of Vietnam share the casualty aversion of political leaders; more junior officers with recent experience of peace operations, on the other hand, realize that full-force protection can be a serious impediment to mission success."[58]

Perhaps the greatest casualty phobia exists among those on the top of the military hierarchy. It appears that "senior military leaders exhibit an intolerance for casualties that far exceeds the intolerance level of the public and policy makers in typical post-cold-war interventions";[59] this pattern appears to be especially acute among the U.S. Army's military leadership.[60] Several explanations account for this acute concern about casualty aversion among the senior military officers:

> For one, officers certainly feel a special responsibility for their troops' welfare. Second, senior officers may lack confidence in the reliability of civilian leaders; thus they fear that the government will abandon the military if casualties mount. Finally, casualty aversion may be an aspect of a growing zero-defect mentality among senior officers, in which casualties are not only deaths—they are an immediate indication that an operation is a failure. If a zero-defect mentality is on the rise, then civilian leaders must share culpability for this problem.[61]

Regardless, self-preservation instincts, or desire to avoid the risks of combat, are decidedly not at the root of this senior military attitude, as those officers most likely to see combat may be the very ones most likely to tolerate casualties.[62]

Considering foreign rather than domestic public opinion, a myth is prevalent that global public attitudes are relatively unified when considering casualty minimization during a particular war or foreign intervention, whereas in reality international opinion on such matters is usually both diffuse and difficult for any one source to influence. Recognizing this predicament, military strategists increasingly think about "structuring a military operation to shape the attitudes, beliefs and perceptions among the enemy and other observers, whether local noncombatants or global audiences";[63] but these efforts have encountered mixed success. If, for example, the United States were to attack an area toward which there existed great international sympathy, or from which the rest of the world received vital natural resources, U.S. policymakers would generally tend to experience great external pressure to implement the most stringent casualty aversion strategies regardless of whether it would be in their best interests to do so. Without such reasons to care, it is equally conceivable that foreign observers might conversely very well pressure the United States to undertake a more brutal campaign so that the conflict could end more quickly.

A closely related myth concerns the constancy of public attitudes about wartime casualties:

> The public's casualty tolerance depends on circumstances that include not only presidential success or failure in mobilizing public opinion but also enemy behavior itself. The Japanese attack on Pearl Harbor instantly dissolved the America First movement as a domestic political obstacle to President Franklin Roosevelt's foreign policy, and the manifest personal and political evil of Saddam Hussein greatly facilitated George Bush's successful demonization of the Iraqi dictator. In contrast, not even the Great Communicator, Ronald Reagan, could explain to the American people exactly what US military intervention in Lebanon was all about; nor could Bill Clinton convey to the public and Congress a persuasive reason for invading Haiti. Unfortunately, although study after study supports the contingent nature of the public's tolerance of casualties, such studies seem to make no impression upon the White House and Pentagon.[64]

Of course, what is most critical here is how the observing public perceives the behavior of the enemy, not the actual enemy behavior. The circumstantial nature of public concern about casualty aversion does, however, open the door to outside manipulation: "Public attitudes toward casualties are malleable, not rigid. Saddam Hussein's repeated miscalculations during the Gulf crisis stemmed in large measure from his twin convictions that Americans could not stand the sight of their own blood and that he was in a position to spill enough of it to collapse US domestic political support for war against Iraq."[65] Regardless, the American public's cost-benefit analysis of other issues—including perceived benefits of a military action, prospects for success, prospective and actual costs, changing expectations, and political leadership—ultimately may be more pivotal than casualty aversion in determining support for military action overseas.[66]

But when exactly is the public most tolerant of casualties? Many point to the justifications for a war as the explanation:

> Bloody, inconclusive wars fought for unconvincing interests are the worst cases for sustaining public support. Conversely, the public is prepared to accept great sacrifice in blood and treasure in situations where vital interests are directly threatened, as they were by the Japanese attack on Pearl Harbor in 1941 and the terrorist attack on the World Trade Center and the Pentagon 60 years later. Most circumstances, however, fall somewhere between outright attack on Americans and the gratuitous and misguided interventions in Lebanon and Somalia, and it is in these middling circumstances—e.g., the Korean, Vietnam, and Gulf wars—where presidential leadership (or lack of it) can make the difference in sustaining public support.[67]

The notion of proportionality in the just war literature,[68] claiming that the overall good achieved by the use of force must be greater than the harm done, generally supports this claim.

The role of political leadership emerges as perhaps the most important influence on public concern about casualty aversion. Several analysts contend that "presidential leadership and the conclusiveness of combat may be more important determinants of public tolerance of casualties than the presence of vital strategic interests";[69] and "elected civilian leaders play a critical role in shaping the public's response to casualties as well as in shaping the characterization of the missions for which such casualties may be incurred."[70] As a result, policy uncertainty and disunity among leaders can lower casualty tolerance.[71]

History of Casualty Aversion

Force protection is of course an ancient an honored military tradition: "An army may avoid decisive engagement as part of an overall strategy designed to exhaust an enemy (Fabius vs. Hannibal), or to bide time while building up one's strength (the Allies against the Axis in the early years of World War II)."⁷² The history of warfare, however, does not reflect an uninterrupted concern with loss of life among either one's troops or innocent bystanders. In early centuries, the conduct of warfare primarily revolved around mercenaries (due to financial savings and the age-old escape route of "plausible deniability") rather than government military troops, and thus there was little concern—or ability to control—the exercise of restraint in the battlefield: governments began authorizing private security forces as early as the thirteenth century, when privateering emerged for the first time; large private armies were widespread in the fourteenth and fifteenth centuries; and mercenaries were commonplace in the eighteenth century.⁷³ After 1800, war generally followed the Napoleonic model, with the objective being one of annihilation—"destruction of the enemy's army in the field and occupation of his country."⁷⁴ Ulysses S. Grant is famous for following exactly this strategy during the U.S. Civil War, costing his side large numbers of casualties. Later on, however, public attitudes and military strategies changed.

Twentieth-Century Force Protection

Given the bloody record of interstate violence in the twentieth century—84 percent of all military and civilian deaths caused by war since 1700 occurred during that century—it is indeed ironic that during this same period the quest for bloodless war accelerated dramatically.⁷⁵ The record of civilian casualties during this century is particularly dismal. For example, in World War I, the 13 million civilian deaths outnumbered the 8.5 million military deaths, and in World War II about 80,000 Japanese were killed by the atomic bomb the United States dropped on Hiroshima on August 6, 1945; indeed, the humanitarian organization Save the Children reports that "the percentage of civilians killed and wounded as a result of hostilities has risen from five percent of all casualties at the turn of the last century to 65 percent during World War II to 90 percent in more recent conflicts."⁷⁶ Worse still, during this century sometimes the killing of innocent civilians was intentional rather than accidental: "The Nazis killed six million Jews in the Holocaust. Millions of Chinese died in the brutal Japanese conquest of the 1930s and 1940s. Between 1975 and 1979, the Khmer Rouge killed two million people in Cambodia. About 800,000 civilians died in the systematic slaughter of men, women, and children in the Rwanda Civil War in

1994."[77] The deliberate slaughter of innocents during wartime represents in every way the precise antithesis of the quest for bloodless war.

Yet when viewed from a broader perspective, war may be becoming increasingly "less gruesome":

> Although the idea that warfare is becoming less gruesome may seem counterintuitive at first glance, it is generally true. During the last two hundred years, both conventional land and naval combat have grown progressively (though not always steadily) less horrible for their participants in the developed world, thanks to factors such as improved medical care and casualty evacuation, mechanization, and refinements in some classes of weapons. Air warfare, too, has become a far less bloody activity over its 90 years of development. In short, the lives of soldiers have, on the whole, become less nasty, brutish, and short since the beginning of the industrial revolution, as have the peacetime lives of civilians. Warfare has also tended to become less brutal for noncombatants, except of course when they are deliberately targeted; particularly in recent years, the ability of armed forces to minimize harm to civilians when attacking their enemies has improved dramatically as a result of the revolution in precision-guided weapons.[78]

At least the potential capacity to conduct war in an orderly and restrained fashion has increased.

World War II, with a staggering 19.4 million people killed in battle, induced foreign security policymakers all over the world to think seriously about the casualty issue. During this most destructive conflict, it was common to believe that "the barbarism of any period pales before the barbarism of today,"[79] and as a result many felt the need to address the issue of loss of life during warfare:

> It isn't ... until after the Second World War that America's statesmen found themselves almost without respite facing the difficult task of objectifying the value of a foreign policy goal in terms of potential losses. There are several reasons for the emergence of this phenomenon. First, the United States was the leader of the Free World and the only power capable of countering the Soviets. International responsibilities rapidly multiplied, and America increasingly found itself engaged militarily around the globe. Second, potential battlefields were often remote and isolated. In the absence of the loss of the fleet at Pearl Harbor, or the sinking of US merchant vessels in the Atlantic, politicians found it exacting to muster and sustain public backing for campaigns abroad. Third, the advent of the nuclear era reduced the utility of total war. Limited war begets limited objectives, which beget more urgent political concerns over casualties. Fourth, the conscription of

the armed forces, reinstituted just prior to America's entry into World War II, was continued until the end of the Vietnam War. Up to a point, a democratic people and their representatives are probably more solicitous about the fate of a draft than a volunteer army. Fifth, and last, America's unrivaled technological prowess offered its leaders at least the hope of achieving victory through science, innovation, and wonder weapons. In the language of an economist, America had a comparative advantage in capital, and was at a comparative disadvantage in labor. With equipment and technology relatively cheap and manpower dear, both economically and politically, it followed that personnel losses would come to be considered as increasingly expensive.[80]

This combination of ingredients gave casualty aversion a decided boost.

The emergence of the Cold War between the United States and the Soviet Union after the end of World War II fostered a proclivity toward casualty aversion, at least on the macro level. Specifically, the Cold War dynamics created a certain cautious stability and restraint due to symmetric sensitivity to casualties from a superpower confrontation: "The fear of suffering massive civilian casualties made both Moscow and Washington careful about even the remote possibility of nuclear exchange."[81] However, on a more micro level, extreme casualty tolerance was evident in many cases—to the point that little official concern about loss of life was ever expressed by the superpowers—during the many limited proxy wars that occurred all over the third world during this era.

U.S. Casualty Aversion

Casualty aversion has very deep roots in the U.S. historical tradition:

> A strong aversion to casualties is rooted in American history and culture. Americans value the individual much more than they do the state, and they have always sought—and with considerable success it might be added—to substitute technology for blood in battle. But only recently has aversion become, at least in the minds of those making war and peace decisions, a phobia—i.e., an aversion so strong as to elevate the safety of American troops above the missions they are assigned to accomplish. Casualty aversion is healthy; casualty phobia is not.[82]

A case in point was White House spokesman Ari Fleischer's observation during Operation Enduring Freedom in Afghanistan in 2001: "I don't think you'll ever witness a nation that has worked so hard to avoid civilian casualties as the United States has."[83] Special U.S. cultural patterns have reinforced the concern about loss of life:

> American society is open and pluralistic, and has historically been served by an independent, skeptical press. Accordingly, US political leaders have been particularly sensitive to the relationship between the cost of war and public support. Even during the United States' first experience with unlimited war, President Lincoln remained concerned in the run-up to the 1864 election that the continuing huge losses of Union troops, in the absence of any clear gains on the battlefield, might lead a weary electorate to abandon him and the cause.[84]

Furthermore, for the United States, "past glory, pride in our current military superiority—which offsets failings in other realms—and a military that consciously presents itself as a vehicle for social mobility and equality all create a heightened sensitivity to the fortunes of individual soldiers."[85] In some ways, "we are the only country that has the strength, wealth and technology to protect not only its citizens from enemies, but also its soldiers in combat."[86] Lastly, the long-standing predilection of the U.S. government to attempt to occupy the moral high ground has driven its quest for bloodless war.

The enduring pattern of healthy distrust by U.S. citizens for government also has played a role here:

> Culturally, American society has a greater distrust of government than many other societies, a sentiment which has historic roots, but which has deepened since Vietnam and Watergate. This distrust means that deaths "for state purposes" must be shown to be necessary, purposeful, and unavoidable. Together with the belief, encouraged by military and political leaders, that war can be—indeed should be—free of American casualties, this distrust can lead some to argue that any such deaths are the result of government incompetence or deceit. . . . Bad things are happening to some soldiers and veterans—there is pain, illness, and premature death among them—but war, in the classic sense, cannot be the cause: the government must be at fault.[87]

Conversely, of course, high levels of loyalty and patriotism—combined with a popular president—can broaden the tolerance for wartime casualties.

Casualty sensitivity is now largely mainstreamed and appears quite regularly in official U.S. security policy pronouncements:

> Currently U.S. military doctrine is sprinkled with references to casualty sensitivity. For example, the official U.S. Army doctrinal manual

for operations . . . declares that "the American people expect decisive victory and abhor unnecessary casualties." However, the Army has not monopolized this perception. The Joint Chiefs of Staff's *Joint Vision 2010* states "the American people still expect us to win any engagement, but they will also expect us to be more efficient in protecting lives and resources while accomplishing the mission successfully."[88]

Indeed, casualty aversion has become a central feature of "the new American way of war":

> Spurred by dramatic advances in information technology, the U.S. military has adopted a new style of warfare that eschews the bloody slogging matches of old. It seeks a quick victory with minimal casualties on both sides. Its hallmarks are speed, maneuver, flexibility, and surprise. It is heavily reliant on precision firepower, special forces, and psychological operations.[89]

In the eyes of many U.S. security policymakers, this kind of war without extensive carnage or long-term commitment fits perfectly the current array of threats.

Recent U.S. strategic doctrine has especially reinforced the emphasis on casualty sensitivity. Government policy appears to rest heavily on what has been informally dubbed the "Weinberger-Powell Doctrine": based on strategic lessons that many military professionals drew from the Vietnam experience, President Ronald Reagan's secretary of defense, Caspar Weinberger, proposed six principles for using force, later amended by George W. Bush's secretary of state, General Colin Powell, who emphasized overwhelming force. These principles are: the presence of vital interests, a determination to win, the establishment of clear political and military objectives, the use of properly sized forces, an assurance of public and congressional support prior to involvement, and the exhaustion of all diplomatic alternatives prior to using force as a last resort.[90] However, this strategic doctrine embodies key contradictions:

> The Weinberger-Powell Doctrine implicitly assumes that public tolerance of casualties is minimal in circumstances that do not satisfy the doctrine's use-of-force criteria, and this assumption elevates casualty minimization above mission accomplishment. Yet, this assumption not only runs afoul of substantial evidence to the contrary but also ignores the role of presidential leadership in shaping public opinion on behalf of using force. The assumption furthermore subverts the integrity of military intervention by compromising its potential operational and strategic effectiveness.[91]

These tensions vividly highlight the more general trade-off often experienced between force protection and mission effectiveness.

Indeed, the casualty sensitivity within the United States is in many ways a source of special security concern, an Achilles' heel that others can exploit:

> Our vulnerabilities, such as they are, lie not in the quantity or quality of our conventional forces nor in weapons of mass destruction. The United States cannot be outresearched, outproduced, or outgunned. Our troops are superbly equipped, brave, and well trained. No one's forces can see more or communicate better in the fog of war than ours. We have great redundancy in nearly every aspect of our forces and economy. And as the Cold War demonstrated, not even the threat to use nuclear, biological, or chemical weapons can stop us from protecting what our leaders manage to define as our vital national interests.
>
> Rather, our vulnerabilities lie within ourselves and our society. Our strength is complimented [*sic*] by our good fortune. We live in a quiet part of the world surrounded by oceans and neighbors who have not the slightest thought of attacking us. As we have come to recognize both our strength and our security, we have imposed constraints on ourselves. In particular, we have grown ever more sensitive about casualties—our own military casualties, opponent and neutral civilian casualties, and even enemy military casualties—and we seek to avoid them. This limits our ability to exercise the tremendous power we possess and makes us susceptible to pressures others can ignore.[92]

In some ways, then, this casualty sensitivity can negate U.S. economic and geopolitical advantages.

Turning to the past chronicle of the U.S. quest for bloodless war, the recent origins go back over thirty years to U.S. involvement in the Vietnam War:

> It derives from America's disastrous experience in Vietnam and prevails among the present national political and military elites, who may have wrongly convinced themselves that the American people have no stomach for casualties, regardless of the circumstances in which they are incurred. Indeed, for these elites, Vietnam is the great foreign-policy referent experience—one seemingly validated by failed US intervention in Lebanon and Somalia.[93]

These apprehensions mistakenly derived from the past appear to be relatively unshakable. The use of the draft during the Vietnam War may have enhanced this casualty concern:

After Vietnam, some analysts traced public concern over casualties to the observation that draftees did most of the fighting. The All-Volunteer Force policy that ended the draft was the politicians' way of avoiding the heat of calling to serve those 19-year-olds who would not want to fight the next war. The Total Force concept, which requires the mobilization of 40-year-old reservists for any serious military engagement by US forces, was the professional's counter-initiative to protect themselves from having to fight alone.[94]

When soldiers feel forced to fight, and neither they nor their families see the point in mandatory conscription, then casualty aversion may escalate.

The contrast between the post-Vietnam attitude and the post–World War II attitude about casualties is remarkable:

> Vietnam produced a generation of political and military leaders that is much more reluctant to use force, or at least use it effectively, than those for whom Munich and World War II were the great foreign policy exemplars. The message of Munich was the imperative of using force early and decisively against aspiring conquerors; the perceived message of Vietnam is that the risks—both on the battlefield and in the domestic political arena—of using force more often than not outweigh the benefits, especially in cases of prospective interventions in other people's civil wars.[95]

Understanding the underlying roots of this dramatic change between the two periods transmits valuable lessons for security policymakers today.

Conflicts in the 1990s reinforced the expectation for few casualties during warfare. Most analysts agree that "by the 1990s, acutely aware of domestic political constraints and seeking to capitalize on America's overwhelming advantages in military technology, firepower, and strategic mobility, the Bush Administration in the early 1990s articulated a defense strategy for regional conflicts" that emphasized providing "forces with capabilities that minimize the need to trade American lives with tyrants and aggressors who do not care about their own people" and that require "the high-quality personnel and technological edge to win quickly and with minimum casualties."[96]

More specifically, "the 1991 Gulf War created unrealistic expectations for low casualties in warfare, not only among the public but also among the military establishment," raising for example the question of "whether commanders can be ruthless enough to pursue the enemy to the limit in the television age when the stakes are less than national survival."[97] Remarkably, "the Gulf War was also the first war in which casualty minimization became from the start an independent operational objective; both the formulation of war aims and the conduct of military operations were governed by dread among the political and especially military leadership that too many American lives lost would implode public and congressional support for the war."[98]

Of all the U.S. military interventions during the 1990s, the one in Somalia has emerged as the ultimate lightning rod for U.S. casualty sensitivity. In many observers' eyes, the "bloody tangle in Mogadishu" reinforced the "overriding concern to minimise the risks to U.S. soldiers."[99] Indeed, the most common interpretation—which may very well be misguided—is based on "the perception that the public supported the deployment so long as there were no casualties, but once the Rangers were killed, support collapsed, and political elites without public support were forced to withdraw their forces, whether or not it was sound policy to do so."[100] The pattern in Somalia reinforces the general notion that "we may expect the public to be particularly sensitive to casualties in humanitarian interventions because these military actions are freely entered into by their government; in this sense, they are 'wars of choice,' as opposed to 'wars of necessity' that must be fought to preserve national security."[101]

Non-U.S. Casualty Aversion

A major question emerges about whether casualty aversion has become a uniquely U.S. phenomenon. The quest for bloodless war is often primarily associated with the United States, but it is not alone in this aspiration:

> All nations with any degree of responsiveness to their citizens are casualty averse, but wealthy democratic countries have acquired a particularly low political tolerance in this area; their political elites can be said to be risk averse in regard to war casualties. This does not necessarily mean these elites are always averse to war, but it does mean they feel a growing pressure to achieve military aims at very low casualty rates.[102]

An unusual demographic explanation has also emerged to explain the moral repugnance of wartime casualties among most Western countries: according to this view, the declining birth rates in advanced industrial societies make each family more unwilling to contemplate loss of a son or daughter in wartime, and so "present attitudes toward combat losses that derive from the new family demography are powerful" in explaining casualty sensitivity.[103]

Examples abound of casualty sensitivity within advanced industrial societies. With regard to the 2003 war against Iraq, Britain exhibited special apprehensions about casualties:

> Britain is perhaps more tolerant of military casualties, having endured decades of guerrilla fighting in Northern Ireland and the 1982 Falklands War, in which 236 servicemen died.
> But, like the United States, it puts a high priority on minimising its own combat casualties and—particularly in this war, which began with scant public support at home and outright hostility in the Arab world—minimising civilian casualties.[104]

Other countries also have predispositions in the exact same direction:

> Although the influence of casualty aversion in Europe or Canada has not yet been the subject of a study, there are obvious indications of a preoccupation with the matter. Lieutenant-General Romeo Dallaire attests to this fact in *Human in Command*. He gives the example of the Belgian contingent which, after having lost ten soldiers in a brush with the enemy in Rwanda, withdrew several days later. Dallaire stresses that the political impact of incurring casualties risks becoming a dominating factor in a commander's decision-making mechanism. He ponders the impact casualty aversion could have on the military ethos of priorities: "the mission, my troops, and then myself."[105]

When facing the Balkan crisis during the 1990s, "Britain and France (not to mention the other putative great power, Germany) flatly refused to risk their ground troops in combat to resist aggression in the former Yugoslavia . . . [I]n the face of atrocities not seen since the Second World War . . . no European government was any more willing than the U.S. government to risk its soldiers in combat."[106]

There may, however, be some differences between U.S. and European attitudes in this regard: "Americans in and out of uniform generally do not consider those who serve as anything other than citizens of equal or more value as any enemy combatant or noncombatant," while Europeans "have a history of populating their militaries with long-term professionals and legions of hired foreigners; this seems to create a certain sense that the armed forces are more tools of the state than citizenry in uniform."[107] Regardless of this difference, many leaders of advanced industrial societies believe that their publics are casualty-intolerant, and for this reason the United States cannot easily contemplate the option of shifting the casualty burden to its allies.

More broadly, an even sharper distinction may exist in attitudes toward the death and destruction of war between the West and the third world: "[I]n the West, a rising aversion to casualties already is shaping national strategies, limiting the use of force to situations that pose little risk to friendly forces," while "in the developing world, warrior cultures glorify war, swelling the ranks of millenarian, fundamentalist, or anarchist movements with thousands of untrained or lightly armed volunteers."[108] Indeed, parts of the third world have such high birth rates and youth bulge that they may be tempted to export underemployed or troublesome young men for battlefield death elsewhere. These developed-underdeveloped cultural patterns appear not only to be in complete opposition to each other but also to be headed in completely opposite directions, spelling considerable trouble for harmonious security cooperation between the North and the South. When military confrontations do emerge between advanced industrial societies and developing countries, a marked asymmetry in casualty sensitivity—potentially disadvantageous to the West—may characterize the violent conflict.

Post–September 11 Force Protection

Given the recency of the terrorist attacks on the United States on September 11, 2001, sharply conflicting opinions surround how this major event has changed attitudes in the United States and overseas about the quest for bloodless war. One pragmatic form of this question, asked with great frequency in recent months, is: When will the ongoing deaths of U.S. soldiers after the wars in Afghanistan and Iraq lead to an outburst of public protest—due not only to casualty sensitivity but also to skyrocketing costs and seemingly unending involvement?

Some feel that the pattern of casualty aversion so consistently evident in U.S. military actions overseas has changed since the terrorist attacks on the World Trade Center and the Pentagon: "America's no-casualties mindset was finally jettisoned in the wake of the September 11, 2001 attacks on New York and Washington. America's leaders decided that the country was ready to support military action that risked sending body bags home."[109] Those subscribing to this view believe the United States has moved away from the zero-casualties mind-set due to fear:

Call it post-Vietnam syndrome, or "casualty aversion"—it has contributed to hastened withdrawals in operations ranging from the victory in the Persian Gulf war to the aborted relief mission to Somalia.

For the moment, such timidity lies buried in the ruins of lower Manhattan and the Pentagon. With perhaps 5,000 American civilians dead from a terrorist attack, military losses suddenly seem like a palatable alternative.

"Our political leaders now understand they are leading a public that fears," said retired Col. Don M. Snider, a professor of political science at the U.S. Military Academy. "The fear is palpable, and when people fear, they are willing to sacrifice to eliminate that fear."[110]

In a *Time*/CNN poll taken shortly after September 11, 2001, of 1,082 American adults, 55 percent favored a ground invasion "that would result in the loss of U.S. lives," and more than 80 percent said they supported air strikes and assassinations.[111] During the 2003 war against Iraq, similar results emerged: "Unlike the previous administration, the Bush team has emphasized that victory in Iraq may involve high human costs," and "the public is getting this message"; a *Washington Post*/ABC poll showed "support for the war remained at 70 percent even though a majority (54 percent) now believed the United States and its allies will sustain 'significant' casualties in the war."[112] Previously, casualty aversion had been fueled by the perceived lack of vital interests or the absence of U.S. elite enthusiasm for foreign military action; however, since September 11, 2001, people may have begun to worry more about threats to their own personal safety and thus seem more tolerant of loss of military lives in the service of their protection.

Underlying this perspective is an emphasis on anger in the United States:

> If anything the pressure is to get MORE involved, even if it puts more US lives at risk. So I do think we've put the casualty phobia behind us to some extent. . . . Given that the terrorist attack on the World Trade Center and Pentagon was a direct intrusion into the United States homeland, it is not surprising that the attitudes associated with the response would be somewhat different—the Sept. 11 attacks provoked a sense of anger and determination among Americans that will surely far outweigh any tendency toward casualty aversion on the part of U.S. society as a whole.[113]

Supporters of this perspective point to the fact that continued deaths of U.S. soldiers in Iraq after the fall of Hussein in April 2003 have not as yet led to overwhelming outcry and political fallout at home. In addition, the change in the country's leadership may also be a factor: "The Bush White House is considerably less casualty-averse than the Clinton White House—that is the single biggest difference; the military is somewhat less casualty-averse now, because prior military casualty aversion was partly a function of doubts about political leadership—Bush resolve has eased those doubts somewhat, but not entirely."[114]

At the very least, in the war against terrorism in Afghanistan, the ground fighting in Operation Anaconda "may have triggered a debate within the administration over the importance of casualty minimization at the cost of mission effectiveness."[115] A Pentagon adviser said that the terrorist attacks on the United States of September 11, 2001, "raised the bar a lot" on the public's acceptance of combat casualties, "but still the administration was concerned about how high the acceptance would be."[116]

In direct contrast, however, some analysts feel strongly that the pre-existing level of casualty aversion in the United States has not diminished. One argument here is that "the leadership's willingness to tolerate American casualties is directly proportional to the stakes, just like for the public; I do not detect a significant change in this post-9/11."[117] Since Pentagon officials has worried about major terrorist attacks on the United States for quite some time, it seems reasonable that they would not change their perspective based on an occurrence that did not dramatically jolt their frame of reference.

Still others assert that since 2001 the tendency toward casualty aversion has been reinforced rather than jettisoned:

> If small wars within failing or failed states have dominated demands on US military power since the Cold War's demise, and especially since the attacks of 11 September, a profound aversion, bordering on the phobic, to incurring American casualties has come to dominate use-of-force decisionmaking in the United States. . . . US combat operations in Afghanistan were . . . conducted in a manner consistent with those of casualty-phobic Operations Deliberate Force (Bosnia) and Allied Force (Kosovo). Either the political and military leadership remained casualty-phobic, or circumstances permitted a cheap victory, or—most probably—both. . . . Clearly, the Bush Administration did not wish to risk a repetition of the Soviet experience in Afghanistan, which mandated avoidance of even the appearance of an overbearing American military presence on the ground. It also had at its disposal an indigenous anti-Taliban alliance inside Afghanistan. Yet reliance on proxies proved a two-edged sword. It spared US lives, but it may also have made it easier for Osama bin Laden and much of the al Qaeda leadership to escape. US troops were withheld from most of the potentially deadly cave searching in the Tora Bora area, leading some commentators

to conclude that unabated casualty phobia—war on the cheap—may have spared bin Laden to fight another day.[118]

Casualty aversion according to this view persisted both during the war on terrorism in Afghanistan in 2001–2002 and during the war against Iraq in 2003. When initiating the war on terrorism in Afghanistan, President George W. Bush refrained "from asking Americans to make any major sacrifices to wage it."[119] And right before the 2003 war against Iraq, U.S. defense secretary Donald Rumsfeld said, "American forces plan to do 'everything humanly possible' and will go to great lengths 'to target only legitimate military targets.'"[120]

Conclusion

Taking into account the broad idealistic humanitarian and practical expedient motivations supporting casualty aversion, buttressed by the perceived casualty sensitivity of the public in Western countries, it is easy to understand why political leaders have seen value in vocalizing the quest for bloodless war to their citizenry. Casualty aversion can receive approval from both high-minded and more pragmatic elements of society, and is difficult to question or oppose in principle under any circumstances. Due to deep Western roots, the quest for bloodless war has become something of a sacred cow in its seemingly universal appeal.

Reviewing the long and checkered history of casualty aversion concerns, it appears that at some level the concerted effort to minimize loss of life during wartime will likely persist and flourish in the future. Especially for the United States, there is "no doubt we will invest much in the effort to keep open the option of bloodlessly imposing our will on others."[121] What is not so clear at this point is whether and when this emphasis serves to undercut other values surrounding mission accomplishment during wartime. Will the trade-offs between casualty aversion and these other values become more intense or relaxed? We must look at more pieces of this complex puzzle to address such issues.

Notes

1. Colonel Trevor N. Dupuy, *The Evolution of Weapons and Warfare* (Indianapolis, IN: Bobbs-Merrill Company, 1980), p. 307.
2. Ibid., p. 310.

3. Major Robert F. Wendel, "Casualty Aversion in the Post-Cold War Era: Defined and Analyzed Through the Logic of Clausewitz," unpublished paper, April 12, 2001, p. ix.

4. John T. Correll, "Casualties," *Air Force Magazine* 86 (June 2003): 50.

5. Project on the Means of Intervention, "Understanding Collateral Damage" (Washington, DC: Harvard University, John F. Kennedy School of Government, Carr Center for Human Rights Policy, June 4–5, 2002), p. 2.

6. Correll, "Casualties," p. 50.

7. Harvey M. Sapolsky and Jeremy Shapiro, "Casualties, Technology, and America's Future Wars," *Parameters* (summer 1996): 119–127.

8. Ben Wisner, "Notes on the Ideas of 'Clean War' and 'Collateral Damage,'" unpublished paper, March 17, 2003, p. 1.

9. John Mueller, "Public Opinion as a Constraint on U.S. Foreign Policy: Assessing the Perceived Value of American and Foreign Lives," paper prepared for presentation at the annual meeting of the International Studies Association, Los Angeles, March 14–18, 2000, p. 12.

10. Daniel Byman and Matthew Waxman, *The Dynamics of Coercion: American Foreign Policy and the Limits of Military Might* (New York: Cambridge University Press, 2002), pp. 137–138.

11. Robert Mandel, "What Are We Protecting?" *Armed Forces & Society* 22 (spring 1996): 335–355.

12. James Turner Johnson, *Morality and Contemporary Warfare* (New Haven: Yale University Press, 1999), p. 5.

13. Ibid., pp. 5, 36.

14. Michael Walzer, *Just and Unjust Wars: A Moral Argument with Historical Illustrations* (New York: Basic Books, 1977), p. 135.

15. Alexander B. Downes, "Targeting Civilians in War: Does Regime Type Matter?" paper prepared for presentation at the annual meeting of the International Studies Association, Portland, OR, February 26–March 1, 2003, p. 2.

16. Lieutenant Colonel Eric A. Ash, "Casualty-Aversion Doctrine?" (summer 2000), available online at www.airpower.maxwell.af.mil/airchronicles/apj/apj00/sum00/ed-sum00.htm.

17. Downes, "Targeting Civilians in War," p. 3.

18. Scott Sigmund Gartner and Gary M. Segura, "War, Casualties, and Public Opinion," *Journal of Conflict Resolution* 42 (June 1998): 279.

19. Mueller, "Public Opinion as a Constraint," p. 2.

20. Gil Merom, *How Democracies Lose Small Wars: State, Society, and the Failures of France in Algeria, Israel in Lebanon, and the United States in Vietnam* (New York: Cambridge University Press, 2003), pp. 19–21, 230–231.

21. David Greenberg, "Fighting Fair: The Laws of War and How They Grew" (January 17, 2002), available online at http://slate.msn.com/id/2060816.

22. Colonel Charles J. Dunlap Jr., "Law and Military Interventions: Preserving Humanitarian Values in Twenty-First-Century Conflicts" (Washington, DC: Harvard University, John F. Kennedy School of Government, Carr Center for Human Rights Policy, Humanitarian Challenges in Military Intervention Conference, November 29, 2001), p. 2.

23. Alex Salkever, "Time to Rewrite the Rules of War?" *Business Week Online* (April 1, 2003), available at www.asia.businessweek.com/technology/content/apr2003/tc2003041_2114_tc047.htm.
24. Dunlap, "Law and Military Interventions," p. 5.
25. Joe Havely, "Why States Go to Cyber-War" (February 16, 2000), available online at http://news.bbc.co.uk/1/hi/sci/tech/642867.stm.
26. Robert Mandel, *The Changing Face of National Security: A Conceptual Analysis* (Westport, CT: Greenwood Press, 1994), chap. 1.
27. Correll, "Casualties," p. 53.
28. Max Boot, "The New American Way of War," *Foreign Affairs* 82 (July–August 2003): 53.
29. Philip Dine, "Practical Politics Guide Effort to Reduce Iraqi Civilian Casualties," *St. Louis Post-Dispatch,* March 16, 2003, p. A8.
30. Brad Knickerbocker, "War Aim: Quest to Reduce Accidental Casualties," *Christian Science Monitor,* March 14, 2003, p. 3.
31. Peter D. Feaver and Christopher Gelpi, *Choosing Your Battles: American Civil-Military Relations and the Use of Force* (Princeton: Princeton University Press, 2004), pp. 151–152.
32. Correll, "Casualties," p. 48.
33. Major Daniel R. Rocha, "Will Unmanned Aerial Vehicles (UAVs) Replace Manned Aircraft?" (1997), available online at www.globalsecurity.org/military/library/report/1997/rocha.htm.
34. Dunlap, "Law and Military Interventions," p. 4.
35. Knickerbocker, "War Aim," p. 3.
36. Sapolsky and Shapiro, "Casualties, Technology, and America's Future Wars."
37. Greenberg, "Fighting Fair."
38. Stephen Biddle, "Land Warfare: Theory and Practice," in John Baylis, James Wirtz, Eliot Cohen, and Colin S. Gray, eds., *Strategy in the Contemporary World* (New York: Oxford University Press, 2002), p. 107.
39. Feaver and Gelpi, *Choosing Your Battles,* p. 153.
40. Harvey Sapolsky, personal correspondence, July 17, 2003.
41. Dupuy, *Evolution of Weapons and Warfare,* p. 311.
42. Tom Bowman, "War Casualties Could Test Public's Resolve: Officials Fear Support Could Shrink As Troops Search for Bin Laden," *Baltimore Sun,* November 18, 2001, p. 19A.
43. Lutz Unterseher, "Interventionism Reconsidered: Reconciling Military Action with Political Stability" (September 1999), available online at www.ciaonet.org/wps/un101/index.html.
44. Ibid.
45. Piers Robinson, *The CNN Effect: The Myth of News, Foreign Policy, and Intervention* (New York: Routledge, 2002), pp. 39–40.
46. Feaver and Gelpi, *Choosing Your Battles,* p. 152.
47. J. F. C. Fuller, *Armament and History: The Influence of Armament on History from the Dawn of Classical Warfare to the End of the Second World War* (New York: Da Capo Press, 1998), pp. x, xiii.

48. Paul Musgrave, "Will Technology Be Used to Make War More Humane? Warfare and the Information Revolution" (February 2003), available online at www.paulmusgrave.net.

49. Max Boot, *The Savage Wars of Peace: Small Wars and the Rise of American Power* (New York: Basic Books, 2002), p. 328.

50. Major Charles K. Hyde, "Casualty Aversion: Implications for Policy Makers and Senior Military Officers," *Aerospace Power Journal* 14 (summer 2000): 18–19.

51. Feaver and Gelpi, *Choosing Your Battles,* p. 7; and Peter D. Feaver and Christopher Gelpi, "A Look at Casualty Aversion: How Many Deaths Are Acceptable? A Surprising Answer," *Washington Post,* November 7, 1999, p. B3.

52. Feaver and Gelpi, *Choosing Your Battles,* pp. 7, 97; Eric V. Larson, *Casualties and Consensus: The Historical Role of Casualties in Domestic Support for U.S. Military Operations* (Santa Monica, CA: RAND, 1996), pp. 1–126; and Jeffrey Record, "Collapsed Countries, Casualty Dread, and the New American Way of War," *Parameters* (summer 2002): 12.

53. Hyde, "Casualty Aversion," pp. 19–20, 22.

54. James Burk, "Public Support for Peacekeeping in Lebanon and Somalia: Assessing the Casualties Hypothesis," *Political Science Quarterly* 114 (spring 1999): 77.

55. Boot, *Savage Wars of Peace,* p. 328.

56. Steven Kull and I. M. Destler, *Misreading the Public: The Myth of a New Isolationism* (Washington, DC: Brookings Institution, 1999), p. 106; and Benjamin C. Schwartz, *Casualties, Public Opinion, and U.S. Military Intervention: Implications for U.S. Regional Deterrence Strategies* (Santa Monica, CA: RAND, 1994).

57. Boot, *Savage Wars of Peace,* p. 328.

58. Theo Farrell, "Humanitarian Intervention and Peace Operations," in Baylis et al., *Strategy in the Contemporary World,* p. 302.

59. Hyde, "Casualty Aversion," pp. 24–25.

60. Record, "Collapsed Countries," p. 12.

61. Feaver and Gelpi, "A Look at Casualty Aversion," p. B3.

62. Feaver and Gelpi, *Choosing Your Battles,* pp. 8, 176–177.

63. Steven Metz, "Strategic Asymmetry," *Military Review* 81 (July–August 2001): 30.

64. Jeffrey Record, "Force-Protection Fetishism: Sources, Consequences, and (?) Solutions," *Aerospace Power Journal* (summer 2000): 7.

65. Ibid., p. 7.

66. Larson, *Casualties and Consensus,* pp. 1–126.

67. Record, "Collapsed Countries," p. 12.

68. Johnson, *Morality and Contemporary Warfare,* p. 28.

69. Jeffrey Record, *Failed States and Casualty Phobia: Implications for Force Structure and Technology Choices,* Occasional Paper no. 18 (Montgomery, AL: Center for Strategy and Technology, Air War College, September 2000), p. 12.

70. Richard A. Lacquement Jr., "Understanding the Casualty Aversion Assertion: Implications and Applications," paper prepared for presentation at

the annual meeting of the International Studies Association, Portland, OR, February 26–March 1, 2003, p. 14.

71. Farrell, "Humanitarian Intervention and Peace Operations," p. 297.

72. John N. Sims Jr., "Shackled by Perceptions: America's Desire for Bloodless Intervention," thesis presented to the faculty of the School of Advanced Airpower Studies, Maxwell Air Force Base, AL, June 1997, p. 28.

73. Janice E. Thomson, *Mercenaries, Pirates, and Sovereigns: State-Building and Extraterritorial Violence in Early Modern Europe* (Princeton: Princeton University Press, 1994), p. 21.

74. Correll, "Casualties," p. 49.

75. William Eckhardt, "Civilian Deaths in Wartime," *Bulletin of Peace Proposals* 20 (1989): 90.

76. Correll, "Casualties," pp. 51–52.

77. Ibid., p. 51.

78. Karl P. Mueller, "Politics, Death, and Morality in US Foreign Policy," *Aerospace Power Journal* (summer 2000): 13.

79. Fuller, *Armament and History,* p. ix.

80. Karl W. Eikenberry, "Take No Casualties," *Parameters* (summer 1996): 113.

81. Byman and Waxman, *Dynamics of Coercion,* p. 222.

82. Record, "Collapsed Countries," p. 12.

83. Dunlap, "Law and Military Interventions," p. 19.

84. Eikenberry, "Take No Casualties," p. 113.

85. Sapolsky and Shapiro, "Casualties, Technology, and America's Future Wars," p. 123.

86. Admiral Gene LaRocque, "Approaching the Digital Battlefield" (Washington, DC: Center for Defense Information, December 15, 1996), available online at www.cdi.org/adm/transcripts/1014.

87. Sapolsky and Shapiro, "Casualties, Technology, and America's Future Wars," p. 124.

88. Sims, "Shackled by Perceptions," p. 77.

89. Boot, "New American Way of War," p. 42.

90. Record, "Force-Protection Fetishism," p. 6.

91. Ibid., pp. 6–7.

92. Sapolsky and Shapiro, "Casualties, Technology, and America's Future Wars," p. 119.

93. Record, "Force-Protection Fetishism," p. 5.

94. Sapolsky and Shapiro, "Casualties, Technology, and America's Future Wars," p. 123.

95. Record, "Collapsed Countries," p. 12.

96. Eikenberry, "Take No Casualties," p. 114.

97. John Chalmers, "Casualty Aversion is Saddam's Greatest Weapon" (March 30, 2003), available online at http://uk.news.yahoo.com/030330/80/dwlff.html.

98. Record, "Collapsed Countries," p. 4.

99. Chalmers, "Casualty Aversion."

100. Burk, "Public Support for Peacekeeping," p. 63.

101. Farrell, "Humanitarian Intervention and Peace Operations," p. 296.

102. Charles Knight, Lutz Unterseher, and Carl Conetta, "Reflections on Information War, Casualty Aversion, and Military Research and Development After the Gulf War and the Demise of the Soviet Union," 1992, available online at www.comw.org/pda/0003refl.html.

103. Edward N. Luttwak, "Where Are the Great Powers? At Home with the Kids," *Foreign Affairs* 73 (July–August 1994): 25.

104. Chalmers, "Casualty Aversion."

105. Lieutenant-General R. A. Dallaire, "Command Experiences in Rwanda," in Carol Pigeau and Ross McCann, eds., *Human in Command* (New York: Kluwer Academic/Plenum, 2000), pp. 37–38; and Colonel Alain Boyer, "Leadership and the Kosovo Air Campaign" (fall 2002), available online at www.journal.dnd.ca/engraph/v3n3_lead1_e.asp.

106. Luttwak, "Where Are the Great Powers?" p. 24.

107. Dunlap, "Law and Military Interventions," p. 17.

108. James J. Wirtz, "A New Agenda for Security and Strategy?" in Baylis et al., *Strategy in the Contemporary World,* p. 314.

109. Ibid.

110. Dan Fesperman, "'Casualty Aversion' Overcome by Terror: Many Americans Prepared to Accept Military Loss of Life," *Center for Defense Information Weekly Defense Monitor* 6 (March 14, 2002): 1.

111. Ibid.

112. Peter D. Feaver, "Casualties Are the First Truth of War—And One the Public Is Well Prepared to Accept," *Weekly Standard* 8 (April 7, 2003): 17–18.

113. Max Boot, personal correspondence, June 11, 2003.

114. Peter D. Feaver, personal correspondence, June 23, 2003.

115. Stephen Biddle, "The New Way of War? Debating the Kosovo Model," *Foreign Affairs* 81 (May–June 2002): 144.

116. Bowman, "War Casualties," p. 19A.

117. Anonymous Department of Defense official, personal correspondence, July 3, 2003.

118. Record, "Collapsed Countries," pp. 10–11.

119. Biddle, "New Way of War?" p. 144.

120. Dine, "Practical Politics," p. A8.

121. Sapolsky and Shapiro, "Casualties, Technology, and America's Future Wars," p. 122.

2

Means of Pursuing Bloodless War

A variety of means exist to pursue the quest for bloodless war. It is not in any way obvious at the outset what is the best approach, or what should be the dominant strategy, in striving for casualty aversion. Indeed, the following discussion of the various offensive and defensive options for casualty-minimizing coercion shows that each approach encompasses very different kinds of assets and liabilities.

Means of Pursuing Bloodless War

Analysts tend to think about the means of casualty aversion in differing ways—some focus on the type of target or initiator, others on the form of weapon, and still others on the nature of the military confrontation. With these differences in mind, a review of available options suggests that four major clusters of approaches have a potential to minimize casualties: banning destructive military action, limiting warfare participants, minimizing civilian exposure to harm, and preventing attack initiation. Some of these means are quite feasible and widely used, while others are relatively infeasible and quite rare.

An overarching approach undertaken in recent years by the United States has been the substitution of matériel for manpower:

> Our sensitivity to our own military casualties is long-standing and certainly not unique. People of all countries love their children and their soldiers, but only we in the United States have the opportunity, the wealth, and the technology to protect them, even in battle. One way to

protect personnel is to remove them as much as possible from the violent edge of the battlefield. US forces have traditionally attempted this by substituting materiel for manpower in war—capital for labor in economic parlance. Today, capital-intensive warfare means soldiers in protective vests or heavily armored vehicles, increasing ranges of fire, and most important, perhaps, vastly increased firepower which attempts to destroy the enemy before many Americans are engaged. Over time we have become quite good at this. In each of our wars we have increased dramatically the tonnage of ordnance delivered per combat soldier exposed. We also have improved significantly the accuracy of our weapons, their reaction times, and the overall pace of our operations. In fact, by now the very worst place on earth to be is in front of an engaged US armored division or on the receiving end of one of our bombing campaigns.[1]

The military advantage of the United States over its adversaries thus does not rest on the number of soldiers that can be thrown into battle but rather on its nearly boundless economic capacity to develop and deliver to a combat zone huge quantities of war-fighting equipment—with superiority in both offensive firepower and defensive protection—that is lethal for enemy targets as well as protective for its own soldiers.

As a result, for the United States there is no need that its primary method of determining victory be who prevails in direct violent human-to-human combat. The age-old "cannon fodder" approach to warfare, in which one overwhelms opponents through vast numbers of soldiers fighting for one's cause, is clearly on the wane if not completely dead, at least among advanced industrial societies. Safeguarding the life of individual soldiers thus may rest more on moral concerns that the expedient need to keep them alive to win the battle. Although the U.S. substitution of matériel for manpower makes great economic sense and contributes to casualty minimization, it is clearly an option that other countries with other combinations of assets would not always be able to select in military confrontations.

Banning Destructive Military Action

This cluster encompasses the most idealistic set of approaches to casualty aversion. Two distinct possibilities emerge here—either inducing disarmament to reduce the devastation resulting from conflict, or making the rules of war more stringent and enhancing their enforcement to lessen the carnage that occurs during warfare. In both cases, the underlying logic seeks to make violent conflicts more civilized, while minimizing their intrusive impact on the ability of a society to function.

However, key obstacles immediately rise to the surface. The first approach violates what appears to be an inexorable pattern across human history of accelerated weapons development and bloody warfare, and the second approach cannot function in an anarchic international system devoid of shared global norms and values on how war should be conducted. Although possible ways certainly exist to move in these directions, such as increased sanctions against genocide or increased efforts to dismantle weapons of mass destruction, generally the feasibility of using this cluster of strategies to minimize loss of life is extremely low.

Moreover, if the motivation behind disarmament is a humane protection of human life, this seemingly moral strategy can backfire:

> The option of not using weapons, lethal or otherwise, provides little solace. Many argue that we should apply economic sanctions and encourage others to join us in diplomatic efforts to change tyrannical behavior or deter international aggression. This policy was applied to communist regimes during the Cold War and to Serbia, Iraq, and Haiti afterwards. The ability of elites to pass on the costs of sanctions to their poor gives pause, however. Many suffer under sanctions, but rarely the intended. We can kill militarily and economically without achieving desired results.[2]

Coercion without weaponry can thus be far more inhumane than coercion with weaponry, as, for example, a food embargo during a conflict can cause the most helpless members of a target population to suffer and die horribly. Moreover, looking back in history before the development of advanced weapons technologies, the conflicts that inevitably emerged in a world without sophisticated arms often involved more protracted violence than those with these arms due to the difficulty of achieving decisive outcomes. Finally, if disarmament initiatives resulted in more equal distribution of coercive capabilities among many states—as do many arms control initiatives—the uncertainty about the balance of power could lead to even more bloodshed.

Furthermore, attempts to tighten the rules of war face huge definitional ambiguities emerging from the rapid pace of change in both weapons development and battlefield tactics. With the increasingly unconventional nature of modern conflicts—involving confusion about the identity of the initiator, the existence of aggression, and the legitimacy of claims—it would be difficult to specify the rules of war in such a way that would effectively discriminate among the range of today's military confrontations even if a consensus were to develop about the restrained conduct of warfare. Even the simplest determination of what kinds of violent turmoil would be properly subject to such rules could be a major challenge.

Limiting Warfare Participants

In recent decades Western powers have evidenced a pattern of wanting to use force overseas without jeopardizing the lives of many of their own citizenry in battle. In some cases, this pattern begins to resemble that of a spectator sport, where—rooted in the fear of broader war and accelerated by technological advances—most of an initiating society does not directly participate in war but rather views it from a distance like spectators at a sports event.[3] Although certainly in line with casualty aversion values for one's own people, this can result in a state becoming numbed to the horrors of a war it has initiated.

Here the approaches depend very much on developments currently just emerging as realistic possibilities. Two strategies under this umbrella are: (1) implementing automated warfare, involving heavily mechanized or robotic military units; and (2) hiring foreigners as private soldiers, altering who fights in wars so that one's own citizenry is not involved. Although both options have been batted around for a long time, government defense establishments are just beginning to consider them as serious casualty aversion strategies, and thus they are in their infancy in terms of sound assessment of costs and benefits in this specific regard. Nonetheless, it is clear that at this point both exhibit only moderate feasibility, albeit with high potential to protect allied civilians and soldiers.

However, both techniques could end up dramatically escalating the damage done to enemy civilians and soldiers, as there would be little incentive for self-restraint. This casualty minimization deficiency could be particularly problematic in a couple of situations: wartime missions whose goals are humanitarian, where for example one is trying to provide food or other supplies to displaced people and provide an orderly basis for their survival (as opposed to trying to overcome enemy soldiers in battle by wiping out their strongholds); and wartime missions whose goals are simply regime change or decapitation of leadership rather than total annihilation of the enemy. In this type of predicament, critical military sensitivity to foreign cultural traditions may be decidedly absent when utilizing foreign private conscripts or automated robots.

Minimizing Civilian Exposure to Harm

This cluster of approaches represents a key impetus—along with increasing battlefield effectiveness, of course—behind recent advances in weapons technology. Although many general missile defense and strategic defense initiatives attempt to protect entire countries—including the armed forces themselves—from exposure to harm, a two-pronged focus has emerged specifically to minimize the devastation of innocent civilian populations: (1) nonlethal weapons that generally incapacitate or stun rather than kill targets (thus usually preventing any unintentionally hit civilians from suffering permanent damage); and (2) precision-guided munitions that can target vital military installations (and avoid civilians and their property) from long distances with incredible accuracy, keeping both those launching an attack and those near the attack zone largely out of harm's way. Distinguishing between acceptable and unacceptable targets, such as between threatening military installations and innocuous civilian social service facilities such as schools and hospitals, has almost always been a part of military strategy, but the primary limitation in the past has been the absence of technological capacity to identify and hit the right targets and avoid the wrong ones.

Thanks to huge recent advances in precision-guided munitions and nonlethal weaponry, the feasibility of both strategies—given adequate intelligence—is substantial, with a high potential to avert casualties among allied civilians and soldiers, and a moderate potential to avert casualties among enemy civilians and soldiers (both technologies can occasionally inadvertently kill or injure people in the target area). However, both distinguishing between target types and protecting nearby citizens and installations have most recently been hampered by the tendency of some enemy leaders to place purposely legitimate military targets right next to protected civilian areas, and the increasing difficulty—even with quality electronic intelligence—to distinguish combatants from noncombatants, and threatening facilities from protected facilities, when the adversary is an amorphous and dispersed nonstate group such as a terrorist organization.

There are, of course, other ways to minimize civilian exposure to harm. For example, insulating onlookers and nearby property assets from wartime collateral damage has been a long-standing tradition, with evacuation of people and movement of valued items out of harm's way the most common approaches. Protective devices could also shield nearby onlookers (and property) from being affected by the devastation associated with warfare. Along these lines, it is even possible to imagine the deployment of regular armed forces to help protect innocent civilians from intentional or unintentional harm during warfare. However, as the lethality of weaponry used in battles has grown and the location of potential attacks has become more dispersed and unpredictable, the ability of such methods to protect civilians—without exception and in all circumstances—has decreased, and these ideas have thus seemed less fruitful.

Preventing Attack Initiation

Preventing attack initiation appears on the surface to be the ultimate cluster of casualty aversion means, as, in a sense, it has the potential to forestall the underlying rationale for violent warfare in the first place. This cluster incorporates preventing adversaries both from launching assaults (negating their offensive capabilities) and from executing coherent counterattacks (negating their defensive capabilities). Here again there are two possibilities: (1) incapacitating or modifying enemy information systems through the use of disruptive techniques and psychological operations that interfere with or alter a target's command-and-control capabilities over its own armed forces (in place of taking enemy troops out directly); and (2) strengthening military deterrence by increasing one's military capabilities and resolve (and possibly shows-of-force) to such a degree that one's superior capacity and will to inflict damage is absolutely unambiguous and credible to all potential adversaries.

The first strategy of information warfare holds great promise, given the extreme dependence of most countries' military operations on electronic computerized communication and information systems and the susceptibility of these systems to alteration and breakdown. The current sophisticated level of propaganda and database penetration systems outstrips the capacity of those with vulnerable systems to protect them, and so the feasibility of this approach is quite high. If accomplished successfully, the potential to save allied and enemy civilian and military lives is quite significant, for states could be paralyzed in terms of their ability to wage war. A difficulty here to be overcome, in light of casualty minimization concerns, is the possibility that, in interfering with military communication and information systems, one inadvertently disables such systems necessary for the survival of the civilian population: for example, disabling an electrical power plant to incapacitate a military installation could leave much of the neighboring population not only powerless but unsafe.

In contrast, the second strategy of deterrence faces many more feasibility problems in confronting the kinds of zealous passionate enemies faced today in the form of both rogue states and terrorist groups who care little about facing death due to overwhelming odds, do not perform careful cost-benefit analyses before launching attacks, and overall seem to thrive on asymmetric warfare. More specifically, because much recent conflict "does not fit neatly into assumptions of rational calculation and instead major elements of irrational 'daredevil' thinking, deterrence seems destined to decline (in effectiveness as well as use) both as a broad strategic doctrine and as a specific means to prevent or end conflict."[4] Although maintaining, enhancing, and demonstrating military capabilities and resolve are absolutely essential prerequisites for any successful coercive strategy, this approach in and of itself is clearly insufficient to signal to potential foes the kind of message that would prevent the occurrence of most of the unruly behavior witnessed today in international relations.

Furthermore, casualty aversion may impede the potential for deterrence because minimizing loss of life projects an image of weakness, lack of resolve, and tentativeness about sustaining commitments to the end. Given the prevalence of asymmetric warfare, enemies with far lower tangible elements of power would probably be downright gleeful about this sensitivity to casualties, which might embolden rather than restrain them in their pursuit of nefarious objectives. If weak adversaries know that they can induce the United States to withdraw from foreign military action simply by instigating casualties among its armed forces, "then they are unlikely to be deterred by U.S. threats to intervene."[5] When a great power intent on minimizing casualties tries to influence them or tell them what to do or not to do, their response would be—according to this logic—downright defiance. Even though enemy challenges to the United States premised on its casualty aversion may involve miscalculation, with an unanticipated U.S. response placing foes on a much steeper escalation ladder than originally envisioned, they can still derive advantages from this casualty sensitivity. A state's willingness or unwillingness to sacrifice human life in the pursuit of foreign security objectives continues to be a crucial signal of credibility in a global age dominated by sometimes-empty democratic rhetoric.

However, closer scrutiny reveals that on at least some occasions the quest for bloodless war could induce an initiating state to use every means at its disposal to build up its powerful image in the eyes of its enemies so that foes would capitulate without much loss of life. This "muscle-flexing" might involve, for example, showing targets that resistance is futile because one's military technology is so superior to theirs that any action on their part can be stymied before it is even launched. Making foes painfully aware of the awesome capabilities of specific casualty-minimizing weapons technologies—such as precision-guided munitions, nonlethal weaponry, and information warfare—can help to create an atmosphere of intimidation. For success in this regard, it would appear that initiators would need to possess an image of credible coercion derived from the past, and in any ongoing confrontation demonstrate superior force early in a campaign.

Moreover, it is at least conceptually possible that some new means may develop in the future to allow a state to communicate superiority of overall power and the futility of resistance in some limited circumstances even to highly passionate or irrational targets. Ideally, of course, the international spread of moral education could make deterrence more effective without substantial loss of life. More practically, however, to cope with the widespread sense that to induce others to change behavior there is no need for massive loss of life, and that, instead, a reliance on more precise, subtle forms of influence will suffice, another approach might have some merit: a more nuanced signaling system—one in which the short-term costs to society may appear to be significantly lower—could emerge in such a way as to salvage deterrence. However, it remains to be seen whether the essential prerequisites are satisfied for a less blunt mode of communication and force demonstration to have a chance to be effective across the wide range of threats confronting the world today.

Trade-offs Associated with Casualty Aversion

Despite the general push in the same direction and surface complementarities among these clusters of casualty aversion means, some critical trade-offs are apparent among them. Pursuing disarmament may undercut any ability to employ precision-guided munitions. Expanding and enforcing the rules of war goes in a direction very different from hiring foreigners—who may not agree with or even be familiar with the rules of engagement of the country employing them—as private soldiers for use in international confrontations. Relying on largely automated warfare but yet engaging in a war where both sides have the ability to incapacitate each other's crucial command-and-control information systems can be an exercise in futility. Strengthening military deterrence while engaging in unilateral disarmament seems virtually impossible. These examples serve to illustrate that engaging in warfare while attempting to use a combination of strategies that minimize the loss of human life may not be as easy as it seems, and thus in such situations those in charge must take care to make sure all the prongs of a military effort are mutually supportive.

Nonetheless, at least the most prominent and commonly used means of pursuing the quest for bloodless war—precision-guided munitions, nonlethal weaponry, and information warfare—generally appear to be mutually supportive and often tightly interconnected. A few examples of the extensive cross-linkages help to illustrate this claim: nonlethal weaponry could serve as part of information warfare to disrupt an electronic command-and-control system; precision-guided munitions could help direct nonlethal technologies to their designated targets; and psychological operations as part of an information warfare campaign could help to magnify enemy forces' images of the destructive impact of precision-guided munitions so as to increase the chances that these foes will lay down their arms.

Outside the trade-offs evident among casualty aversion strategies, one issue insufficiently acknowledged when considering the means of pursuing the quest for bloodless war is that inescapable trade-offs often exist between casualty-minimizing and war-winning (often casualty maximizing) strategies. In other words, it is not frequently the case that the best way to prevent loss of life during a war is also the best way to achieve victory. An obvious example is when the avoidance of certain highly lethal military strategies, out of concern for preserving human life (soldier or civilian, friend or foe), ends up preventing a decisive military outcome and lengthening the confrontation. Because dramatic inconsistencies exist between casualty-minimizing and casualty-maximizing strategies, an integrated war-fighting approach combining both may pose a real challenge. The opportunity costs of choosing to have casualty sensitivity play a key role—at times a dominant role—in a country's military doctrine are thus quite substantial.

Any coercive strategy has to by necessity balance a number of competing objectives:

> Warfare is about balancing three goals. On the one hand, you must accomplish military objectives, like seizing territory or destroying enemy forces. On the other hand, you must accomplish political objectives, the larger geopolitical goals that the combat is meant to serve, like stability in the region. On the third hand, you must bring back alive as many of your soldiers as possible.[6]

It is readily apparent that, if the protection of forces decreases the ability to attain important objectives, there is a real problem. When and if force protection communicates timidity, lack of resolve, or unwillingness to commit for the long haul on the part of the initiator, this value can clearly undercut key goals and cause a military campaign to fail.

The competing values exhibited by casualty-minimizing and casualty-maximizing strategies in part reflect the manner in which statesmen and generals come at the bloodless war issue from very different vantage points:

> Statesmen and generals consider battle losses from different perspectives. The former must weigh the repercussions of excessive casualties on the level of civilian morale necessary to successfully prosecute a war, and ultimately (at least in a representative form of government) on their own political futures. The latter, on the other hand, must balance potential losses against a wide variety of military factors including probable strategic or tactical gains, possible damage to the effectiveness of the forces employed and their ability to cope with enemy countermoves, and the difficulty of reinforcing or reconstituting the force. Military commanders, when planning and conducting operations, must also respond to their civilian leaders' guidance (if any) concerning the number of casualties deemed politically acceptable.[7]

Of the two sets of considerations, it appears that generals have the more difficult task, for while statesmen primarily judge the impact of casualties on political support for the war, generals—in addition to responding to political leaders' guidelines—must not only take into account the traditional military cost-benefit analysis but also gauge the effect of casualties on their own troops' determination to fight in future battles. Moreover, military commanders in the field often form a personal bond with soldiers around them—something rarely experienced by political leaders back home—further complicating generals' calculations.

Even for statesmen, however, deciding where to draw the line when it comes to choosing a military strategy that determines what level of human sacrifice to tolerate during wartime is not always easy:

> Political leaders who consider committing their nations to war, or sustaining their country's ongoing participation in an armed conflict, must consider how many lives they are willing to see expended in order to accomplish the objective. They must concern themselves both with the public's attitude to personnel losses, and to the expected reaction of the political opposition. While there are numerous variables influencing these attitudes and reactions (for instance, the openness of the political system, the role of the media, and the economic costs of a war), a central factor in all cases is the degree to which involvement in a conflict is regarded as a legitimate defense of a nation's vital interests.[8]

It is interesting to note that, within democracies, politicians' concern about casualties may thus come less from their own personal convictions and more from their worries about adverse reactions from political supporters and opponents. Deciding how many lives one is willing to sacrifice based on this kind of political calculus, or on a determination of the extent to which a predicament threatens vital national interests, is extremely subjective and difficult to justify.

When the zeal for casualty aversion causes political leaders to focus as a panacea on just one of the values that need balancing, there is little doubt that putting all one's eggs in the basket of high-technology warfare has its drawbacks. More specifically, in light of the common direct trade-offs with key military objectives during warfare, an overemphasis on force protection as an operational imperative can certainly have some unintended and even bizarre side effects. For example, "it can result in the concentration of force when security for aid operations would best be promoted through the dispersal of peace forces to provide military presence over a larger area; and it can require military commanders to order their forces to wear body armour, visibly demonstrating distrust and insecurity (as US forces did in Somalia), when a more relaxed force posture would make it easier to build relations with the local communities (as British forces did in Sierra Leone)."[9]

Outside of mission accomplishment issues, other trade-offs surround the means of pursuing bloodless war. A key trade-off may exist between force protection and the safety of noncombatants. For example, with limited military resources, commanders in the field may have to choose in deploying them whether to protect their soldiers or protect civilians; or in the heat of battle, commanders may have to make tough decisions about using a weapon that would decimate the enemy—and thus shield their troops from possible harm—yet also kill numerous innocent bystanders. Furthermore, a narrow focus on casualty minimization can often serve to oversimplify an assault strategy:

> The history of warfare provides overwhelming evidence that high-tech "smart" weapons are but one dimension of combat power. For instance, Serbian militia can counter the most sophisticated gadgetry the United States can throw against them simply by positioning their mortars next to mosques. The American military must always plan for tough foes like the North Vietnamese, resilient against the most modern arsenal available, and count the incompetent armies of Noriega and Saddam Hussein as windfalls. War is still a complex political act, not a computer simulation, and combat is still an art, not a science. Hills remain to be taken by soldiers fighting their way up to the summit. Too enamored of the technical solution, the United States may produce the wrong mix of equipment and structure its forces inappropriately.[10]

With an overly constrained set of combat tactics, even the most astute military commanders would have a difficult time persevering and winning on the battlefield.

Having casualty aversion interfere with mission objectives is particularly problematic today given the frequently marginal national interests associated with post–Cold War great power involvement in foreign military confrontations. Strong Western governments no longer see that it is in their self-interest to intervene to achieve stability in distant parts of the world, with an increasingly unclear basis for legitimate coercive action due to the uncertain payoff, the high risks of involvement, and murkiness about which side to assist. Many of the opportunities for foreign war or intervention necessitate a kind of long-term low-intensity involvement in which even the strongest of these nations' militaries have rarely been particularly successful. The pervasive indifference to such foreign predicaments among these powerful states' mass publics, whose wealth and position make them want protection for themselves (despite their questionable loyalty to the state) but at the same time make them reluctant to sacrifice their lives for the protection of others in faraway lands, reinforces this noninvolvement.[11] Overemphasized casualty minimization in this context can often precipitate pressures for quick withdrawal before mission objectives can be decisively accomplished.

Future Means of Casualty Aversion

Out of the large range of approaches discussed, three specific strategies that serve to minimize casualties currently demonstrate the most promise—precision-guided munitions, nonlethal weaponry, and information warfare—and these receive extensive separate analysis in Chapters 3–5. In the future, however, some different, less orthodox means of pursuing bloodless war may emerge more prominently. In the desire to minimize human casualties, one increasingly appealing focus is to consider alternatives to using one's own citizens or engaging in killing during warfare. Rather than using new technologies to augment the actions undertaken by one's own citizens fighting as soldiers abroad, these emerging nontraditional methods involve developing a sizable substitute for one's own human fighting forces. Two such transformational strategies stand out—utilizing largely mechanized or robotic armed forces or private foreign armed forces.

Largely Mechanized or Robotic Armed Forces

There may be a move toward largely mechanized warfare involving the use of robotic soldiers and unmanned vehicles to prevent the need to put human lives in peril. In either case, advancing technology, not force protection, becomes the primary mission. Aside from self-defense, in the future it is clear to many that, if casualty aversion concerns predominate, "only such conflict as can take place without soldiers is likely to be tolerated . . . and robotic weapons will be used increasingly."[12]

Much discussion has emerged recently detailing possible uses of robots in warfare:

> Once science fiction, today the robots and the attack laser are fact. They are part of a massive research effort in labs across the US that will give the US military the ability to dominate the battlefield of the 21st century. . . . Critics say that the US already has the most powerful armed forces in the world and will retain that position far into the future. But the US military believes technologically superior weapons will reduce casualties and deter future wars if an enemy believes it has no hope of defeating such a formidable force. . . . On the future battlefield, robots will be king. The US Air Force envisions unmanned aircraft that can be launched from submarines, ships or runways, and has already awarded Lockheed Martin a $120 million US contract to start work on such a project. An unmanned aircraft can fly faster and higher at half the price of manned fighters. The US Navy envisions

using robot crabs to scurry across beaches, defusing mines. The US Army wants to use robots for battlefield reconnaissance. In October, the Marines will test a robot mortar called Dragon Fire; once the device's sensors detect an enemy formation the mortar begins firing. The Marines are also considering using the robot mortar in conjunction with drone planes that could locate an enemy.[13]

The specter of combat in future wars could thus be dramatically different from the past:

> Within 20 years, squadrons of unmanned planes will swarm enemy sites like killer bees, launching missiles and avoiding detection with sophisticated jamming devices.
> Self-programmed submarines will replace dolphins to detect and disarm mines. Robotic mules the size of pickups will haul ammunition, medical supplies and food.
> Drone ambulances will load wounded soldiers and cart them to hospitals. Crablike robots will crawl into buildings to sniff out chemical stashes.
> The transition to mechanized weaponry is key to the military's transformation from heavy ground forces to smaller human units fortified with robotic weapons. The goal: to limit casualties.[14]

One of the challenges embedded here is that developers appear, so far, to be better at conceiving new ideas for automated weapons than at developing specific practical scenarios in which they could be integrated in the field with traditional combat tactics to achieve military objectives with minimum loss of life.

There also has already been significant forward movement in the area of unmanned assault vehicles:

> Additional money should be moved into technologies that reduce air crew exposure to possible loss. This means investment in ever longer-range stand-off precision munitions, and, above all, in unmanned aerial vehicles (UAVs). The greater the precision, the smaller the munition required, and therefore the greater the reduction of impact on people and things not intentionally targeted.
> Increased reliance on UAVs (which include cruise missiles) and on large stand-off platforms like the B-2 may encounter resistance among the so-called "fighter mafia" which dominates the Air Force leadership. ... Yet UAVs and the B-2 satisfy the imperatives of elite casualty phobia. Both are difficult to target; the UAVs in any event don't have aircrew; and while B-2s do, the ratio of crew to capacity to deliver precision munitions swamps that of any "tactical" aircraft. ...

Of particular priority is the need for an autonomously-piloted vehicle. Such a vehicle has been advocated for years, but the technology required is still out of reach, as is even some of the science. A UAV that is remotely controlled has great limitations because of the situational awareness of the "pilot," and to obtain that awareness would require a large data "pipe" that could be vulnerable and almost certainly would greatly increase the cost of the UAV. An autonomous vehicle solves these problems, yet introduces others because it would require a computer that could "think" and that would be asked to make life-and-death decisions on the battlefield.[15]

The unmanned aircraft Predator, for example, proved to be vital in the war against terrorism in Afghanistan in 2001–2002 to find and track targets.[16] Ironically, the use of UAVs may in some ways make foreign military intervention look more like police action than part of a formal war. While extensive mechanization creates major challenges of its own, such as issues of control, breakdown, and repair, it appears to have the potential to advance the bloodless war agenda: "Although UAVs have been around for the last 40 years, they have only recently been aggressively pursued and developed as vital tools for use in military operations. [Through their use] the US will be able to minimize risks while flying against heavily-defended targets."[17]

Rather than being part of an unrealistic pipedream, policymakers are already taking concrete steps in the direction of automating military force. For example, on July 8, 2003, the U.S. House of Representatives approved by a vote of 399 to 19 a whopping $369.1 billion Pentagon spending bill for 2004, which begins a shift toward the lighter, more mobile combat systems favored by Defense Secretary Donald H. Rumsfeld, and which specifically earmarked $1.7 billion for the U.S. Army's Future Combat System, a network of ground-based and air-based robots, sensors, lightly armored vehicles, and guns.[18] In sharp contrast to skepticism and opposition to this kind of idea in decades past, there appears now to be growing acceptance that robots are an inevitable—and potentially lifesaving—component of fighting forces of the future.

At the moment, however, many analysts view largely mechanized or robotic assaults as useful primarily only in exceptional circumstances as a last resort:

> For the really tight spots, we are developing robots to replace soldiers. Already robots guard some installations and can be used to enter dangerous buildings or approach suspected bombs. Soon, no doubt, they will be able to take a share of the real fighting, demonstrating an ability to

identify and engage resisting targets on a smoke-obscured battlefield under remote direction.[19]

As the technology gets better and cheaper, security policymakers may consider these avant-garde fighting technologies for a wider range of applications.

Finally, a few critics still worry that mechanized or robotic assaults might increase questionable activities during combat:

> Assassinations, destruction of individual sites, and counter-intelligence missions will be far more common. When flying robots weighing less than a gram can act as spies collecting accurate, digital, and up-to-date information are feasible—as they will be soon—then organizations won't need a full-fledged network of human spies to gather data. Those same robots in slightly different configurations could easily kill a selected individual or group, perhaps by putting strychnine in their coffee or cyanide in their corn flakes.[20]

Others are concerned simply about ineffectiveness—shifting tasks from humans to robots has long been met with limited success because "robots cannot reproduce the complexity of the human brain, as they react poorly to unexpected circumstances, which is what war is all about."[21] Lastly, there is a long-standing worry that the use of heavily mechanized equipment and robots during warfare may lead eventually to removal of humans from the decisionmaking loop, leading to the kind of out-of-control military mayhem depicted so well in the 1983 film *War Games*. However, even those most critical of largely mechanized or robotic armed forces recognize the inevitability of movement in this direction, spurred at least in part by the quest for bloodless war.

Private Armed Forces Composed of Foreigners

The other future possibility of unorthodox casualty aversion is fighting wars using soldiers who are not one's own citizens. In order to circumvent the intolerance of casualties for civilians and soldiers from one's own country, "one scheme would be to follow the Ghurka model, recruiting troops in some suitable region abroad" and having them serve as private mercenaries for the commissioning government in wars overseas.[22] To some analysts, "greater US cultivation of and reliance on local surrogates to assume the risks of ground combat" makes sense because "when they are willing and (with training and assistance) able to fight a common enemy, they limit America's potential military liabilities in circumstances in which domestic political tolerance of US casualties is—or is believed to be—severely limited."²³ This idea also involves the potential for using members of transnational mercenary groups and private military companies.²⁴

Although private armed forces composed of foreigners have been used for centuries, several recent examples serve to show specifically how private foreign forces could be used to prevent casualties among one's own government troops. In Bosnia and Afghanistan, for example, "large indigenous friendly ground forces served as surrogates, thus holding down US casualties and their perceived potential domestic political consequences."²⁵ The role of private military companies later received a significant if little-noted boost, albeit not directly a part of war-fighting strategy, in November 2002, when the Virginia-based contractor DynCorp received a new assignment from the State Department's Diplomatic Security Service to help protect Afghan president Hamid Karzai.²⁶ From 1994 to 2002 the Pentagon entered into more than 3,000 contracts with private military firms, serving to "provide the logistics for every major American military deployment," including a plan to hire a private paramilitary force to guard sites in Iraq after the 2003 war.²⁷

The leaders of the great powers "have become quite terrified of taking casualties" through wars or interventions overseas, and as a result private military forces have begun to look awfully attractive:

> An American ambassador in Europe told dinner guests a couple of years ago that his country could no longer emotionally, psychologically or politically accept body bags coming home in double figures. By the start of the Kosovo war, just 15 months ago, that number had been reduced to zero. So we tried to fight a war from 15,000 feet. That taught us the limits of stand-alone air power. We couldn't stop or slow the pogroms, so we creamed the capital city of the guilty nation until after 74 days a fat Russian stepped in, slapped down his protégé Slobodan Milosevic, and procured a chaotic form of peace. We managed to kill 14 times more Serb civilians than uniformed soldiers and zero secret-police killers. But we avoided casualties and called it a victory. The utter horror of taking casualties has not extended to Britain and France, but is subscribed to by the rest of Europe. As for any kind of involvement in a lethal hellhole in Central or South America, Africa or Asia, simply on humanitarian grounds—forget it. We might use our own troops to extricate our own citizens, or even to protect a massive national economic or strategic facility, but that is about it. We watch

the charnel house of Sierra Leone with horror but impunity. Then into the frame, to politically correct cries of "Yuck," steps the professional mercenary.[28]

The idea of victory without casualties among one's own citizenry can be extremely appealing in democracies with a narrow sense of national interest.

Private sources of military force, whether they are highly unstructured and informal like modern mercenary groups or highly formal and structured like today's major private military companies (such as Virginia-based Military Professional Resources Incorporated), appear to have some special military advantages for minimizing friendly casualties. Mercenaries could feel free to engage in a wide variety of extralegal types of force, including covert assassination, kidnapping, and sabotage. Speed and flexibility are trademarks of private military action, as there is no need to go through layers of bureaucracy to get approval for a mission or go through months of preparation to ensure there will be absolutely no casualties among those undertaking the assignment. While the mass public and international onlookers may under limited conditions care about the deaths of civilians and government soldiers, an average consumer would probably be relatively indifferent to the deaths of unknown mercenaries, and to a certain extent feel free of moral responsibility. Perceived costs could thus potentially be much lower here, while not sacrificing effectiveness, than with government military action.

Such private foreign armies could even be used to combat transnational terrorism.[29] Counterterrorist action is an area in which Western states have been particularly unsuccessful, in terms of both destroying targets and protecting noncombatants from harm. Private military forces appear to be most needed and effective in coercive counterterrorist action when a state government is concerned about and sensitive to force protection among its own official armed forces but yet feels it needs to employ intense coercion exceeding normal bounds against one or more terrorist groups.

Nonetheless, in contemplating the use of private foreign soldiers during wartime, potential problems emerge of both fighting power and control: "Relying on a proxy army can have its drawbacks, especially in a place such as Afghanistan, where ethnic and tribal rivalries run deep and the United States might not be able to control its surrogates."[30] Private armies may also not have access to the most sophisticated war-fighting military technologies. Third world countries may perceive Western reliance on private foreign soldiers for military action abroad as a signal of even less commitment than the use of government troops conveys today. Moreover, when a government chooses to outsource to private security providers, the attraction may result from the state bearing little public accountability for undesired consequences, deaths of citizens, or moral and legal dilemmas surrounding the legitimacy of a foreign military action.[31] Mercenaries have in recent times been less scrupulous about the exercise of restraint during war, including respect for human rights and protection of innocent civilians in war zones.

Conclusion

Within the diverse set of means available to pursue casualty aversion, none is capable alone of maximizing this value. Each strategy aids in one dimension of the protection of human life and of overcoming the adversary, but contains critical deficiencies for handling other critical dimensions present in some types of violent international confrontations. All of the casualty aversion strategies presented have to cope with and overcome a series of major obstacles, and none is equipped to handle them all with ease and alacrity. Critical trade-offs surround the means of pursuing bloodless war, not only among casualty-minimizing strategies but also between these strategies and those war-fighting approaches that maximize casualties in the short run, and between force protection and the safety of noncombatants. Even the techniques of casualty aversion that may be applied in future warfare face similar limitations. Loyal and trained military manpower, placed directly in harm's way and with the full range of coercive strategies at its disposal, still seems essential for many missions.

Compounding limitations on the battlefield is often a lack of in-depth comparative understanding of the strengths and weaknesses of these means of casualty aversion among the military leaders responsible for choosing among them. On the political side, the ambivalence of Western governments about the nature of their international responsibilities in a post–Cold War environment makes guidance to these military leaders often obscure or nonexistent. In comparison to the relatively clear mutual understandings between the two blocs during the Cold War, today's degraded communication system in a global anarchic environment makes it unclear how casualty aversion strategies can work best to signal resolve to the enemy. On the military side, the difficulty of gauging the exact relationship between casualty aversion and mission accomplishment makes pivotal choices tricky indeed.

Thus pursuing the quest for bloodless war, whether justified or unjustified, is not easy. Equally challenging to those selecting which war-fighting means to pursue is the need to balance casualty aversion with other values critical in foreign missions to achieve overall effectiveness. The attempt to accomplish ambitious military objectives through force, while at the same time protecting human life, may in the end—compared to alternatives involving the unrestrained use of force—often be more difficult and costly and more likely to result in disappointment and resentment.

Figure 2.1 summarizes the potential utility of the full set of strategies encompassed by the four clusters of casualty minimization approaches: banning destructive military action, limiting warfare participants, minimizing civilian exposure to harm, and preventing attack initiation. Aside from the overall superiority of precision-guided munitions, nonlethal weaponry, and information warfare as instruments for the quest for bloodless war, some other interesting patterns are evident. First, the most idealistic approaches—initiating disarmament and expanding and

Figure 2.1 Strengths and Weaknesses of Casualty Aversion Means

	Potential Capacity to Protect People on Both Sides of a Conflict				
Casualty Aversion Means	Operational Feasibility	Allied Civilians	Allied Soldiers	Enemy Citizens	Enemy Soldiers
Ban destructive military action					
Induce disarmament	Very low	High	High	High	High
Expand and enforce rules of war	Very low	High	High	High	High
Limit warfare participants					
Hire foreigners as private soldiers	Medium	High	High	Very low	Very low
Implement automated warfare	Medium	High	High	Very low	Very low
Minimize civilian exposure to harm					
Employ nonlethal weaponry	High	High	High	High	Medium
Use precision-guided munitions	High	High	High	High	Medium
Prevent attack initiation					
Incapacitate/modify information systems	High	High	High	High	High
Strengthen military deterrence	Low	High	High	High	High

enforcing the rules of war—would be incredibly well suited to minimize loss of life were it not for their complete infeasibility in today's anarchic international system. Second, in a parallel fashion, the realpolitik approach of deterrence would prevent war from breaking out due to fear of overwhelming retaliation and nicely fit with the value of casualty aversion were it not for the tendency of most of today's most unruly adversaries to ignore any such cost-benefit calculus. Third, both of the approaches that may become more widespread in the future—implementing largely automated warfare and hiring foreigners as private soldiers—have a tremendous potential to protect allied troops from harm but at the same time are likely to increase the carnage inflicted on enemy soldiers and civilians. Thus in facing the challenges of the future we are certainly not bereft of potentially effective means of casualty aversion, but at the same time we do not have that many arrows in our quiver.

Notes

1. Harvey M. Sapolsky and Jeremy Shapiro, "Casualties, Technology, and America's Future Wars" *Parameters* (summer 1996): 119.

2. Ibid., p. 122.

3. Colin McInnes, *Spectator-Sport War: The West and Contemporary Conflict* (Boulder: Lynne Rienner, 2002).

4. Robert Mandel, *The Changing Face of National Security: A Conceptual Analysis* (Westport, CT: Greenwood Press, 1994), pp. 7, 24; and Robert Mandel, *Deadly Transfers and the Global Playground: Transnational Security Threats in a Disorderly World* (Westport, CT: Praeger, 1999), p. 35.

5. Benjamin C. Schwartz, *Casualties, Public Opinion, and U.S. Military Intervention: Implications for U.S. Regional Deterrence Strategies* (Santa Monica, CA: RAND, 1994), p. 4; and Major Charles K. Hyde, "Casualty Aversion: Implications for Policy Makers and Senior Military Officers," *Aerospace Power Journal* 14 (summer 2000): 19, 27.

6. Peter D. Feaver, "Casualties Are the First Truth of War—And One the Public Is Well Prepared to Accept," *Weekly Standard* 8 (April 7, 2003): 17–18.

7. Karl W. Eikenberry, "Take No Casualties," *Parameters* (summer 1996): 109.

8. Ibid.

9. Theo Farrell, "Humanitarian Intervention and Peace Operations," in John Baylis, James Wirtz, Eliot Cohen, and Colin S. Gray, eds., *Strategy in the Contemporary World* (New York: Oxford University Press, 2002), p. 302.

10. Eikenberry, "Take No Casualties," p. 115.

11. Summary of Proceedings, Defense Intelligence Agency Conference, "The Privatization of Security in Sub-Saharan Africa," unpublished document, Washington, DC, July 24, 1998, pp. 1–2.

12. Edward N. Luttwak, "Where Are the Great Powers? At Home with the Kids," *Foreign Affairs* 73 (July–August 1994): 27.

13. David Pugliese, "The Future of War," *Defence Associations National Network News* 5 (summer 1998), available online at www.sfu.ca/~dann/backissues/nn5-2_3.htm.

14. Jon Swartz, "New Breed of Robots, Gizmos Take War to the Next Level," *USA Today,* May 12, 2003, p. 3B.

15. Jeffrey Record, *Failed States and Casualty Phobia: Implications for Force Structure and Technology Choices,* Occasional Paper no. 18 (Montgomery, AL: Center for Strategy and Technology, Air War College, September 2000), pp. 21–22.

16. David A. Fulghum, "UAVs Whet the Appetite," *Aviation Week & Space Technology* 158 (March 3, 2003): 52.

17. Major Daniel R. Rocha, "Will Unmanned Aerial Vehicles (UAVs) Replace Manned Aircraft?" (1997), available online at www.globalsecurity.org/military/library/report/1997/rocha.htm.

18. Dan Morgan, "House Approves $369 Billion for Defense Spending: Funds Lay Groundwork for Shift That Rumsfeld Is Seeking to Lighter, More Mobile Military Force," *Washington Post,* July 9, 2003, p. A04.

19. Sapolsky and Shapiro, "Casualties, Technology, and America's Future Wars," p. 122.

20. Paul Musgrave, "Will Technology Be Used to Make War More Humane? Warfare and the Information Revolution" (February 2003), available online at www.paulmusgrave.net.

21. Swartz, "New Breed of Robots," p. 3B.

22. Luttwak, "Where Are the Great Powers?" p. 28.

23. Jeffrey Record, "Force-Protection Fetishism: Sources, Consequences, and (?) Solutions," *Aerospace Power Journal* (summer 2000): 10.

24. Robert Mandel, *Armies Without States: The Privatization of Security* (Boulder: Lynne Rienner, 2002).

25. Jeffrey Record, "Collapsed Countries, Casualty Dread, and the New American Way of War," *Parameters* (summer 2002): p. 4.

26. David Isenberg, "Security for Sale in Afghanistan," *Asia Times Online* (January 4, 2003), available online at www.atimes.com/atimes/central_asia/ea04ag01.html.

27. P. W. Singer, "Have Guns, Will Travel," *New York Times,* July 21, 2003, p. 15A.

28. Frederick Forsyth, "Send in the Mercenaries," *Wall Street Journal,* May 15, 2000, p. 1.

29. Robert Mandel, "Fighting Fire with Fire: Privatizing Counterterrorism," in Russell D. Howard and Reid L. Sawyer, eds., *Defeating Terrorism: Shaping the New Security Environment* (New York: McGraw-Hill, 2003).

30. Tom Bowman, "War Casualties Could Test Public's Resolve: Officials Fear Support Could Shrink As Troops Search for Bin Laden," *Baltimore Sun,* November 18, 2001, p. 19A.

31. David Shearer, *Private Armies and Military Intervention,* Adelphi Paper no. 316 (London: Oxford University Press, International Institute for Strategic Studies, 1998), pp. 69–72.

3
Precision-Guided Munitions

Of all the means of pursuing the quest for bloodless war, precision-guided munitions (PGMs) have received the greatest publicity. Military strategists have sought to develop precision armaments for a longer period of time than for the other primary instruments—nonlethal weaponry and information warfare—of casualty aversion. Indeed, within the defense establishment today, this approach appears to have attained the widest acceptance.

Given the tangible success with precision weaponry in recent conflicts, many analysts inside and outside the U.S. military are now calling for a restructuring of our coercive capabilities so that the overwhelming emphasis is on precision-guided munitions.[1] Buttressed by the unprecedented accuracy evidenced in the recent wars in Afghanistan and Iraq, many onlookers are proclaiming that the basic nature of violent conflict itself has changed. Combining the humanitarian potential to minimize collateral damage and civilian casualties with the efficiency potential to minimize the number of bombs dropped necessary to hit vital targets, on the surface there appears to be no downside to this development.

The current fascination with precision-guided munitions has striking parallels in earlier eras. For example, a quarter century ago some analysts were so amazed at the capabilities of these "pinpoint" arms that they thought they signaled the "death of the tank."[2] Indeed, traditionally many observers hail each successive increase in accuracy as permanently altering the way the military should operate in future battles. Many recent general claims about the advantages of precision-guided munitions have been dangerously sweeping. The emerging conventional wisdom is that "if smart bombs are good, smarter bombs are better"; and that "high technology munitions are a panacea for conventional vulnerability," assuming that "what can be seen can be hit, and what can be hit can be destroyed."[3] This optimistic assessment influences both conceptual security discussions and more pragmatic evaluation of Pentagon budget requests and demands for reforms in military spending or modernization of military capabilities.[4]

Thus a pressing need exists to question this frequently unqualified set of assertions, and more specifically to begin to isolate when low-destructiveness precision munitions (differentiated from massively destructive imprecise weaponry by their high accuracy and low casualty rate) are most and least effective in today's international conflicts. To do so, this chapter discusses definitional controversies surrounding precision weaponry, including the precision/brute-force trade-off; the motives for utilizing precision weaponry; the recent history of the application of precision-guided munitions; dangers of overreliance on precision weaponry; and the conditional utility of these arms in current foreign conflicts.

Definition of Precision-Guided Munitions

The definition of what exactly constitutes precision weaponry has changed over time as arms technology has evolved. Originally termed "smart bombs," the broader classification of "precision-guided munitions" later became official.[5] Although the meaning of "precision" is decidedly relative to the time in which it is used,[6] the term has always involved amplified accuracy with increased range or distance. In the broadest sense, the September 11, 2001, terrorist attacks on the World Trade Center and the Pentagon were precision-guided, with humans directing the explosives toward the target. Despite the inherent ambiguities, perhaps the most commonly accepted specific notion of a precision-guided munition is "a bomb or missile that is guided during its terminal phase," with a high probability "of making a direct hit at full range—when unopposed—on a tank, ship, radar, bridge or aircraft."[7] Some even utilize a numerical threshold in terms of "circular error probable" (CEP), measuring the radius of a circle within which 50 percent of the projectiles fired will impact, as the basis for isolating precision-guided munitions.

Even with a clearer notion of what constitutes precision, it is difficult to determine how to evaluate the effectiveness of such weaponry during warfare. The many ingredients contributing to the success of military coercion involve assessing the balance of resolve, national interests at stake, the capacity of each side to inflict pain on each other's civilian population, and coercive forces possessed by each side.[8] Most of these thus go well beyond a simple cost-benefit analysis of both whether one can wipe an opponent off the face of the map or whether losses associated with surrender outweigh possible gains associated with continued struggle, and instead involve a sensitive political and psychological assessment of the mentality of the sides at war. At one level effectiveness of precision-guided munitions could rest on achieving the initiator's regional political-military objectives, and at another level such effectiveness could mean satisfying desired target destruction goals; it is certainly possible to be a total success at one of these two aims and a complete failure at the other.

Some inappropriately lump precision-guided munitions with air power, and by implication massive imprecise destruction with ground power, thereby converting the controversy about precision weapon utility into an air-land power debate. This misconception may be amplified by the tendency of the U.S. Air Force to be ahead of the curve when it comes to precision-guided munitions, while the U.S. Army is apparently still struggling to come to terms with the full implications of this technology.[9] Although in recent wars most precision-guided munitions have been delivered through aerial bombardment, this is not intrinsically necessary, and there have been several important launches of precision-guided munitions from ships at sea, from tanks on the ground, and even from weapons held by individual soldiers fighting land battles. In any case, unlike in earlier periods, it is increasingly evident that the particular weapons platform "has become less important, while the quality of what it carries—sensors, munitions, and electronics of all kinds—has become critical"; the sophistication of the vehicle from which the weapon is launched or dropped is less significant than the weapons guidance technology itself.[10]

As weapons technology has advanced through time, some armaments have focused on improved accuracy or precision, while others have emphasized greater obliteration potential or destructive impact. One way to illustrate the common though not inescapable trade-off between these two ideal types of weaponry is to compare two very traditional implements of warfare, the cannon and the rifle. The cannon (particularly the old bombard) has huge destructive power but very low precision, while the rifle (particularly a sniper rifle) has great precision but relatively low aggregate destructive power. Nobody would reasonably suggest that one is universally superior to the other or that one could readily be substituted for the other, and no matter how much the accuracy and range of the rifle were improved it would still face severe limitations in certain kinds of coercive confrontations. More modern highlighting of this distinction occurred recently in Afghanistan and Iraq: compare the surgical accuracy of Tomahawk land attack cruise missiles (equipped with relatively light 1,000-pound homing warheads) or the joint direct attack munitions (JDAMs) with the absolute obliteration potential of the 15,000-pound BLU-82 Daisy Cutter bomb or the massive ordnance air blast bomb (MOAB), a 21,000-pound munition appropriately nicknamed the "Mother of All Bombs." The oft-quoted distinction between the sledgehammer and the well-directed rapier cuts along the same lines. Of course, gray areas persist: for example, thermobaric weapons are precise in that they can land at the mouth of an Afghan cave, yet blind/indiscriminate in that they kill everything inside.

Increased accuracy usually associates with decreased destructive power. This is exemplified even in the most recent application of precision-guided munitions, in the 2003 war against Iraq:

> Paradoxically, increasing precision makes U.S. firepower both more effective and less destructive. Because U.S. bombs can hit within a meter or two of their aim point, they carry a lighter load of explosives. U.S. war planners tried hard to minimize collateral damage by employing the smallest possible munitions to get the job done, on occasion going so far as to drop bombs filled with nothing but concrete.[11]

Thus the trade-off between accuracy and destructiveness is a function of both technology and military objectives.

There appears to be a direct relationship between these two contrasting types of military weapons and two opposing types of military orientations. If your coercive thrust is totally "annihilating the enemy before he annihilates you,"[12] utterly destroying your foe's infrastructure and ability to function, then weapons of massive indiscriminate destruction rather than those involving precision targeting seem more appropriate; if in contrast your coercive thrust consists of surgical strikes against select enemy targets, maintaining the integrity of your foe's infrastructure and ability to function, then precision-guided munitions rather than weapons of massive destruction seem more appropriate. While precision attacks eliminate a few key targets in the hope that the enemy will extrapolate the damage done to damage that might occur later, the annihilation strategy in no way depends on such logic on the part of a foe. A parallel distinction contrasts obliterating an enemy and influencing an enemy:

> Brute force overcomes an obstacle simply by destroying it, as the Romans overcame Carthage in the Third Punic War (149–146 B.C.). Rome did not employ force to influence Carthage to engage in desired behavior; there was nothing Rome wanted Carthage to do. Rome simply wanted to wipe Carthage from the face of the earth. The actual use of brute force, then, can entail the destruction of an opponent. In these situations, influence is not the object. In most situations where force is used, however, influence is the aim. The use of force is usually meant to hurt the opponent until the opponent's will to resist further is broken.[13]

From this perspective, then, minimally destructive high-precision weaponry would associate more with attempting to achieve influence, while maximally destructive low-precision weaponry would associate more with attempting to eradicate one's opponent.

Making this distinction between influence and annihilation is not without controversy. Some analysts might argue that obliterating or annihilating the enemy is not a meaningful choice in today's security environment, and that therefore only the influence end of the continuum receives realistic security consideration. However, this argument seems fundamentally flawed: extreme brutality and strategies of national annihilation "are not characteristics of the distant past only," with, for example, European states practicing extermination in some areas they ruled "well into the twentieth century";[14] and even today a huge range of coercive military action exists, with genocide initiated by states or ruthless groups approaching the annihilation end of the continuum and shows-of-force approximating the opposite influence endpoint. In practice, of course, many wars involve a mix of the two types of weapons and objectives, often implemented sequentially with the targets that need initially to be taken out with high confidence—such as air defense batteries, command-and-control, missile transporters and launchers, and weapons labs—subject to precision attacks first. Nevertheless, understanding the precision/brute-force trade-off, summarized in Figure 3.1, helps to reveal when each is most useful.

Thus the use of precision weaponry depends much more heavily than the use of brute-force weapons on initiators' understanding and utilizing the psychology of the target's interpretation about desired behavior. Precision-guided minimally destructive munitions associate more

Figure 3.1 Brute Force Versus Precision Weaponry

Brute Force	Precision
Initiator's objectives toward target	
Complete devastation of targets	Coercion
Obliterating enemy	Influencing enemy
Massive indiscriminate destruction	Pinpoint bull's-eye accuracy
Annihilation	Select surgical strikes
Initiator's military munitions	
Imprecise maximally destructive weaponry	Precise minimally destructive weaponry
Sledgehammer	Well-directed rapier
Bombard cannon	Sniper rifle
Daisy Cutter bomb	Tomahawk cruise missile
Massive ordnance air blast bomb	Joint direct attack munitions
Initiator's needs regarding target responses	
Interpretation of desired behavior not vital	Interpretation of desired behavior vital
Extrapolation of vulnerability not vital	Extrapolation of vulnerability vital
Anticipation of future damage not vital	Anticipation of future damage vital
Choice about whether to submit not vital	Choice about whether to submit vital

with influence, and imprecise maximally destructive arms associate more with brute force because precision weapons rely for success on enemy calculations, based on the demolition of key targets with great accuracy from great distances, that resistance is futile and surrender is essential; while imprecise maximally destructive weapons do not rely on such extrapolation, or on any form of signal interpretation by the enemy about desired behavior, due to the complete obliteration of the adversary. Only in cases where such interpretation is necessary do target foes have a choice about whether or not to submit,[15] with precision-guided munitions potentially affecting enemies' calculations as to their own vulnerability. One advantage of this influence-oriented approach is, as Sir Basil Henry Liddell Hart reminds us, that today's enemy may be tomorrow's customer or even ally: "To inflict widespread and excessive destruction is to damage one's own future prosperity, and, by sowing the seeds of revenge, to jeopardize one's future security."[16] However, it is important to remember that the distinction between brute force and coercion (or between obliteration and influence) "is not always easy to discern" in the heat of battle.[17]

A key facet of this distinction revolves around a target's anticipation of what is going to happen in the future. As Thomas Schelling appropriately notes, "It is the threat of damage, or of more damage to come, that can make someone yield or comply. . . . [I]t is the expectation of *more* violence that gets the wanted behavior, if the power to hurt can get it at all."[18] In attempts at influence rather than obliteration, following the logic of deterrence and compellance, the goal of the coercer is "the use of enough force to make the threat of future force credible to the adversary."[19] Even considering the complexity of calculations determining reactions from other countries, the use of precision weaponry can contribute to this credible expectation of future force by an initiator against a target.

Motivations Spurring the Growth of Precision Weaponry

The development of precision-guided munitions has substantially affected traditional military thinking. The established tradition has reflected that, "since the campaigns of Generals Ulysses S. Grant and William T. Sherman, the 'American way of war' has called for stocking massive amounts of materiel and supplies in theater for decisive victory."[20] Even today many strategists believe that keeping large volumes of military matériel forward in-theater is a sound military policy. Yet many modern international conflicts require smaller and more precise weaponry. The changing nature of warfare has triggered demand for new armaments:

> In tracing the historical development of PGMs, from the rudimentary radio controlled weapons of the Second World War to the state-of-the-art laser-guided, rocket-boosted munitions of the Vietnam era and beyond, it is clear that these weapons did not emerge simply as a natural consequence of technical change, any more than did earlier missile guidance systems. Rather, American scientists and engineers responded to a very real demand by U.S. policymakers for an "ultimate weapon" that would inflict upon an adversary not mass destruction, but precise, calculated damage.[21]

Moreover, "in addition to developing new technology, the United States has tried to improve the ability of all the military services to use this technology in order to minimize collateral damage and risks to U.S. personnel through changes in doctrine and training."[22]

For the United States in particular, it was the urgent need in the Vietnam conflict that provided the catalyst spurring the development of certain precision-guided munitions, just as the Gulf War triggered the development of the GBU-28 bunker buster. The operational need to increase greatly the lethality of air attacks on bridges was pivotal: bridges have been notoriously difficult targets to destroy, requiring at least several sorties with unguided bombs much of the time, and precision weaponry largely solved that problem. While much of the conceptual basis for precision-guided munitions is not new, "it is only since laser-guided bombs were used in Vietnam that military planners have generally agreed that they were economically and operationally feasible."[23]

Military Motivations

The primary motive for military interest in precision-guided munitions is to improve efficiency and lethality against point targets.[24] Due to their accuracy, initiators can achieve high damage expectancy with few sorties, moving from a multiple-sorties-per-target mode of operation to a multiple-targets-per-sortie mode. Lieutenant-General Joseph Kellogg Jr. argues, considering the recent use of precision weapons in Afghanistan, "We haven't eliminated the fog of war, and we never will . . . but we can sure reduce the thickness of the fog and enable the ability to see through it."[25] The technological advances in precision targeting appear in many ways useful for both "improving the potency of military threats and neutralizing an adversary's countermoves."[26] The utility of precision-guided munitions is enhanced due to the increasing combat dependence today on command-control-and-communications systems (often involving sophisticated computer information retrieval and artificial intelligence) that are highly vulnerable to disruption by this type of weapon.[27] When fully operational, precision-guided munitions may even facilitate attacks against totally unseen foes hidden by distance, terrain barriers, or human-constructed cover.

Political and Psychological Motivations

More complex, however, are the long-term political and psychological premises behind the development of precision-guided munitions. For politicians, "the collateral damage and casualties generated by unguided weapons" pose major problems, and so the "use of PGMs in a conflict shows a political sensitivity and sophistication that is appreciated around the world."²⁸ After the United States developed and deployed them, it discovered that they could help overcome political obstacles to destroying targets that were proximate to "do-not-touch" zones such as those including apartment buildings, schools, hospitals, commercial centers, or cultural symbols. The U.S. concern about collateral damage associated with the development of precision-guided munitions reflects the perceived public distaste for harming innocents:

> In recent decades, technologies have been used both to minimize U.S. casualties and to counter accusations that the United States does not care about adversary civilian suffering. One answer to North Vietnam's attempt to exploit collateral damage was the U.S. introduction of more-advanced precision-guided munitions against targets likely to draw harmful propaganda, such as air defense sites in populated areas. When striking terrorist camps in Afghanistan in 1998, the United States used cruise missiles, in part because they posed no threat to U.S. personnel, even though a manned-flight bombing mission could have inflicted greater damage on the terrorist training camps that the United States sought to destroy.²⁹

Precision weaponry has the potential to reduce harm to both attackers and defenders,³⁰ and to instill "confidence" in policymakers "confronted with having to contemplate using force in circumstances where so-called 'collateral damage' would be either unacceptable or call into question the viability of continued military action."³¹

This worry about collateral damage associated with precision-guided munitions has deep roots. First, there is the basic moral concern about the sanctity of human (noncombatant) life: precision weaponry hurts fewer people than large, maximally destructive dumb bombs; and even when compared to how economic sanctions affect civilians, modern air warfare using precision-guided munitions "stands out as an increasingly efficient, effective, and humane tool of foreign policy."³² Despite increasingly ambiguous civilian-military target distinctions, democracy-induced adherence (at least cosmetic) to global humanitarian norms and concern about perceived adverse domestic and global public opinion have served to underscore the quest for precision. More pragmatically, some advocates of precision-guided munitions fear that too much carnage may rouse up enemy sympathizers, as for example civilian casualty sensitivity could inflame the rest of the Islamic world during the war against terrorism in Afghanistan.

History of Precision-Guided Munitions

Throughout history there has been great effort to identify the attributes of armaments that influence battlefield effectiveness. Military historian Major General J. F. C. Fuller treats accuracy of aim as one of five attributes of weaponry, along with range of action, striking power, volume of fire, and portability.[33] Similarly, based on a review of past uses of arms in warfare, Colonel Trevor Dupuy calculates the killing power of weaponry—what he terms the "theoretical lethality index"—as including such elements as rates of fire, number of potential targets per strike, relative incapacitating effect, effective range, muzzle velocity, reliability, battlefield mobility, radius of action, and vulnerability.[34] The history of weaponry shows that both precision and destructiveness have increased over time, and that militaries have usually wanted and needed both kinds of arms to achieve their objectives in varied kinds of conflict. Thus precision is but one of the ingredients contributing to coercive potential, and not necessarily the most important one at that; yet this is the element that has perhaps the closest relationship with casualty aversion.

Regardless, the quest for precision has been among the most important stimulants to advances in technology. Because "ever since military men began shooting things at enemies, most shots have missed or been ineffective."[35] Over time, defense concerns about accuracy have greatly intensified. "The elusive goal of killing one's opponent from a distance, with absolute precision, has been a holy grail of warfare":[36]

> The TV image, if not the reality, is of weapons entering airshafts of buildings to destroy enemy command and control capabilities. Although some of these weapons "lose lock" and go astray, and although many more sorties still deliver dumb bombs rather than precision-guided munitions, the hope is for silver bullets that are certain to target and that remove the need for sustained attack and for the flattening of an enemy city.[37]

Particularly in the twentieth century, military weapons research has focused on this issue:

> As with many great breakthroughs, precision munitions initially appeared to be merely an extrapolation of existing technologies and not a fundamentally new approach to warfare. . . . [E]ventually, it became clear that there was something unprecedented in this new

class of projectiles, something unheard of since firearms became preeminent some five hundred years ago. Whether rocket- or tube-based, the new projectile was not bound by the laws of ballistics. In theory, it could not miss the target, as it could be adjusted to compensate for the target's evasive maneuvers. In practice, it had a far greater probability of hitting the target than a conventional ballistic weapon. More important by far, there was no theoretical limit as to the distance it could go.[38]

Many saw precision weaponry as a major leap forward due to unprecedented improvements in targeting accuracy (with the maximum error down in some cases to just a few inches) and targeting efficiency, with the capacity to destroy a bridge with a single pass of an aircraft.

Not everyone has been swept along, however, by the precision-guided munitions craze. Perhaps the most famous qualifying response comes from General Tommy Franks, head of U.S. Central Command in Operation Enduring Freedom in Afghanistan: confronting zealous onlookers flush with recent success desiring to stock up on precision weaponry, he stressed that the military must not view that experience as a template for all future conflicts—"one size will not fit all."[39] More generally, some critics argue that "PGMs make it possible for fewer aircraft to destroy more targets than in the past, but this enhanced efficiency makes little difference to the coercive effectiveness" of a wide range of military strategies: "Bombing knocked out nearly all power generation in North Korea (90 percent), North Vietnam (85–90 percent), and Iraq (over 90 percent), but in no case caused the population to rise up against the regime. . . . [I]f modern nation-states can withstand so much, they will not give in under the relatively bloodless harassment envisioned by today's [precision] strategic bombing advocates."[40] Within the military itself, ground forces are particularly skeptical of some computer-based targeting systems: "They wonder whether any technologist can disperse what Carl von Clausewitz called the fog of war and ask what will happen when an opponent attempts to conceal its force or attacks the information systems that observe it."[41] The ongoing offensive-defensive arms development cycle ensures that no particular weapons technology can be permanently dominant.

Across the history of warfare, technological advances in accuracy have on occasion yielded unexpected consequences. It is common to hear that, "contrary to the illusion of precision and calculability conveyed by advanced professional management . . . total war proved to be far more intractable to intelligent decision than managers expected";

instead, considering specifically World War I and World War II, "the marvelous ingenuity of technology and military management" resulted in "unexpected destruction of life and property," producing among many "a deep agony and revulsion in the world."[42] More broadly, "success in coercive contests seldom turns on superior firepower" measured in technological terms; indeed, "the United States has often seemed to have the most difficulty coercing its least powerful foes."[43] Possessing technologically more advanced weaponry, including advantages in precision, may dissuade adversaries from confronting a given country symmetrically, but may actually encourage bold foes to confront the state in armed conflicts that are asymmetrical, engaging it where its superiority is not unquestioned.

In order to illustrate the actual track record of precision-guided munitions in the distinctive context of post–Cold War conflict, four military campaigns stand out as ripe for scrutiny—Operation Desert Storm in Iraq during the 1991 Gulf War, Operation Allied Force in the former Yugoslavia during the 1999 Kosovo conflict, Operation Enduring Freedom in Afghanistan during the 2001–2002 war against terrorism, and Operation Iraqi Freedom in Iraq during 2003. The United States was centrally involved in all three operations, although Allied Force was an initiative of the North Atlantic Treaty Organization (NATO). Each case involved the use of precision-guided munitions, and there is a pattern of increased reliance on this type of weapon across the first three campaigns: in Operation Desert Storm, 118,700 sorties were flown and 265,000 bombs delivered, of which 20,450 were precision-guided (the United States delivered 89 percent of the precision-guided bombs); in Operation Allied Force, 37,500 sorties were flown and 23,000 bombs delivered, of which 8,050 were precision-guided (the United States delivered 80 percent of the precision-guided bombs); and in Operation Enduring Freedom, 38,000 sorties were flown and 22,000 bombs delivered, of which 12,500 were precision-guided (the United States delivered 99 percent of the precision-guided bombs).[44] Many consider Operation Iraqi Freedom to be the most precise war in history, as the dependence on precision weaponry reached unprecedented heights; U.S. defense officials estimated that precision-guided bombs accounted for nearly 70 percent of all bombs used by the coalition during the campaign:[45]

> Not only did the US make nearly ten times as much use of precision guided weapons relative to unguided weapons as in the Gulf War, it was able to target them with far more focus and effect. As for sheer

numbers, nearly 100% of the combat aircraft the US deployed in the Iraq War carried precision weapons, versus some 15% of the aircraft in Desert Storm. The Coalition fired some 19,948 precision-guided weapons in the less than four week long Iraq War versus 8,644 in the six week Gulf War, and some 955 cruise missiles versus 300.[46]

Operation Desert Storm

Operation Desert Storm was a forty-three-day campaign initiated by the United States and its allies against Iraq between January 16, 1991, and February 28, 1991. Its stated goal was to liberate Kuwait, in response to Iraq's invasion in August 1990 and subsequent declaration of Kuwait as an Iraqi province. After coalition aircraft targeted strategic Iraqi military targets such as airports, command-and-control centers, missile launch sites, and radar stations, the primary ground thrust occurred only during the last 100 hours of the war. By the end of February these coalition forces had defeated the Iraqi military and freed Kuwait.

U.S. pilots used precision-guided munitions for a variety of purposes in Operation Desert Storm, despite the far greater cost at the time of guided as compared to unguided munitions: GBU-12 laser-guided bombs destroyed more than 200 tanks a night during the last weeks of the war; GBU-15 electro-optical glide bombs destroyed oil manifolds to stop oil from flowing into the Persian Gulf after Iraqi forces opened the valves; GBU-24 laser-guided bombs destroyed chemical, biological, and nuclear storage areas, bridges, aircraft shelters, and other strategic targets; GBU-27 laser-guided bombs hit hard targets such as aircraft shelters; and AGM-65 Maverick missiles attacked armored targets.[47] With the precision capability of U.S. aircraft, one $70,000 Maverick equated to a $1.5 million T-72 tank, since it only took one missile to destroy each Iraqi tank; General Buster Glosson, planner of the Gulf War air strikes, argued that "two raids of 300 B-17 bombers could not achieve with 3,000 bombs what two F-117s can do with only four.... [O]f the 85,000 tons of bombs used in the Gulf War, only 8,000 tons (less than 10 percent) were PGMs, yet they accounted for 75 percent of the damage."[48] The success of this precision weaponry initiative was due to many elements, including the high level of U.S. intelligence pertaining to Iraqi targets, the rather traditional state-to-state nature of the warfare with formal military targets, the difficulty of mixing civilians and combatants together to deter attack, the ease of identifying enemy vulnerability points, and the absence of an enemy population with a passionate resolve to continue to resist after military defeat was evident.

However, a balanced assessment reveals that special auspicious conditions influenced the success of precision weaponry during the Gulf War. "The Gulf War showed air power off to great advantage but in extremely favorable circumstances—the United States brought to bear a force sized and trained to fight with the Soviet Union in a global war, obtained the backing of almost every major military and financial power, and chose the time and place at which combat would begin in a theater ideally suited to air operations."[49] Even with these advantages, Operation Desert Storm revealed some air-to-surface weaponry shortcomings, with weather conditions limiting employment of precision-guided munitions and smoke, fog, and overcast skies making laser designation virtually impossible.[50] Despite the benefits provided by advanced precision-guided technology, Saddam Hussein remained in power and defiant of the United States and the international community for over twelve years afterward, indicating possible limits on the ability of this technology to achieve long-term political changes.

Operation Allied Force

On March 24, 1999, the North Atlantic Treaty Organization initiated Operation Allied Force as a means to compel Serbian leader Slobodan Milosevic to cease ethnic cleansing and human rights violations in Kosovo, primarily against ethnic Kosovar Albanians, and to pull Serbian forces out of the disputed province. Fourteen nations waged the attack entirely by air, severely damaging Serbian infrastructure before ultimately ending Milosevic's rule. Like Operation Desert Storm, the war was remarkably one-sided, as the U.S. defense budget alone was fifteen times the size of Serbia's entire gross national product.[51] Nonetheless, while Operational Allied Force "revealed once again that precision-guided munitions are essential to high-leverage, low-attrition, low collateral damage attacks; it also revealed a chronic pattern of under-investment in these weapons in the United States but even more so among the allies."[52]

The employment of precision-guided munitions in this campaign was quite extensive. By the end of the war, U.S. and British surface ships had deployed 218 Tomahawk sea-launched cruise missiles, and B-52 bombers had deployed another 90 conventional air-launched cruise missiles: Operation Allied Force involved the first combat use of the satellite-guided joint direct attack munition from altitudes of 40,000 feet delivered by the B-2 bomber; and unlike in Desert Storm, this weapon—along with conventional air-launched cruise missiles, and selected Tomahawk missiles—proved to be impervious to weather.[53] In the end, "NATO air forces (or rather, American air power, modestly augmented by Europe) demonstrated persuasively that precision air weapons can destroy from a safe distance any large fixed target like a bridge, power station, oil refinery or factory."[54]

Although utilizing a greater percentage of smart weapons than in any other conflict before it, the campaign's success required a surprisingly sustained effort over seventy-eight days:

> While we can marvel over the demonstrated capabilities of B-2s, JDAMs, laser-guided munitions, and Global Positioning System-assisted bombing techniques, looking at Allied Force objectively, it still looks like a win, but a rather ugly one. For starters, we got that "W" by applying a greater portion of the Air Force's total airframes during this operation than we did in the Korean War, during any period in the Vietnam War, or in Desert Storm. It took the air forces of 13 contributing NATO countries to batter Yugoslavia to the point that Milosevic agreed to withdraw his forces from Kosovo and permit the introduction of a UN peacekeeping force, including Russian troops, into the strife-torn province.[55]

Despite the precision of NATO air attacks, the estimated 5,000 to 10,000 casualties included hundreds of unintended injuries or deaths in Serbia, the displacement and massacre of large numbers of ethnic Albanian Kosovars, and people hurt through an embarrassing accidental demolition of a wing of the Chinese embassy in Belgrade, all indicating some significant holes in intelligence and communication. The campaign in many ways qualified the utility of precision-guided munitions, which "however useful against enemy armor in the open field, are next to useless in cities and in partisan warfare."[56] The unconventional nature of the warfare, involving nonstate military targets without clear demarcation from civilian installations, certainly complicated matters. Indeed, Serb units intentionally operated in the midst of civilian refugees and near prohibited targets, utilized extensive camouflage, concealment, and deception to fool NATO, and employed high-altitude antiaircraft missiles that forced NATO planes to fly at high altitudes, all making it difficult to definitively identify critical vulnerability points on the ground.[57] Most important, despite the removal of Milosevic, Kosovo today is not characterized by peace, democracy, stability, or prosperity.

Operation Enduring Freedom

Operation Enduring Freedom began on October 7, 2001, four weeks after the September 11 terrorist attacks on the World Trade Center and the Pentagon in the United States. The explicit goal was to defeat the Taliban and Al-Qaida forces, identified as the perpetrators of the violence. There were about 60,000 U.S. soldiers involved in the war effort at the peak of the conflict. A coalition of Afghan Northern Alliance fighters, air sorties by U.S. planes, Western special operations forces and intelligence operatives, and a small contingent of Western ground forces engaged in the action. Although most of the bombing had ended by late 2001, when the new government regime in Afghanistan was inaugurated, sporadic bombing continued through mid-January 2002, and this campaign is still continuing at the time of this writing.

Precision weaponry dominated the battle in Afghanistan as in no other before it:

> By most measures, America's high-tech armaments have performed superbly in Afghanistan. In April 2002, in its preliminary review of the Afghan bombing campaign, the Pentagon found that more than three-fourths of U.S. bombs—smart and dumb alike—hit their intended targets. And the precision of the smart bombs (meaning bombs and missiles guided to their targets by lasers or satellites) was breathtaking. The Navy, for example, claimed a 90 percent target hit rate for its smart bombs. Should those estimates hold, the Afghanistan air offensive would define a level of accuracy never seen before in wartime. Compare those figures with the 1991 Persian Gulf War and the 1999 NATO-coordinated strikes against Yugoslavia, where fewer than half of all allied munitions hit their intended targets.[58]

Satellite-fed data from ground troops guided old Vietnam-vintage B-52s to drop bombs in designated 1,000-yard-long areas, and the U.S. Air Force reports that of 6,650 JDAMs dropped, less than 10 percent missed their targets.[59] Operation Enduring Freedom was in many ways a model of technological efficiency: while in Desert Storm the military flew about 3,000 sorties a day, in Enduring Freedom the number was reduced to 200 sorties a day; General Tommy Franks, head of U.S. Central Command, said the 200 sorties a day hit roughly the same number of targets hit with 3,000 sorties a day in Desert Storm, and that while about ten aircraft were needed to take out a single target in Desert Storm, a single aircraft was used to take out two targets on average in Enduring Freedom.[60] The cost of this precision weaponry had plummeted—while the price of the primary precision weapon in the Gulf War, the Tomahawk cruise missile, was more than $1 million apiece, the price of the primary precision weapon in Afghanistan, the joint direct attack munition, was just $18,000 for a kit that used a global positioning satellite system to convert a dumb bomb into a smart one.[61] The high-quality ground and air intelligence available to air-war planners and combat pilots in Operation Enduring Freedom made it possible to identify many key vulnerability points (even given a dispersed and rather elusive set of targets) and launch precision-guided munitions against them within just ten minutes.[62]

However, this campaign also highlighted limitations of precision-guided munitions. "Despite the adulation of Operation Enduring Freedom (OEF) as a 'finely-tuned' or 'bulls-eye' war, the campaign failed to set a new standard for precision in one important respect—the rate of civilians killed per bomb dropped; in fact, this rate was far higher in the Afghanistan conflict—perhaps four times higher—than in the 1999 Balkans war."[63] In part, this pattern reflected the incredibly high difficulty in distinguishing between civilians and the informally clad terrorist combatants. Moreover, the reported success of precision weaponry may be misleading, as "U.S. airpower has been essentially unopposed,"[64] with the Taliban lacking the know-how to undertake deception or camouflage and the technology to implement electronic jammers and air defense systems.[65] The Al-Qaida terrorists' unbridled passion for their cause and their apparent unwillingness to accept defeat also certainly has made the battle harder. Most important—despite the use of the most precise weaponry available—the war effort has ultimately failed to achieve (at least as of the time of this writing) a key goal of unambiguously capturing or killing Osama bin Laden and other top enemy leaders.

Operation Iraqi Freedom

On March 20, 2003, U.S. and British troops invaded Iraq to depose Saddam Hussein. Despite U.S. claims that its coalition included forty-nine nations, the only other countries to provide fighting forces were Australia and Poland (on land) and Denmark and Spain (at sea). The stated justification for the invasion included allegations that the Iraqi government had produced weapons of mass destruction and engaged in human rights violations, and thus the primary goal was to rid the country of these arms and free the Iraqi people from the oppressive Hussein regime.

The invasion reached a speedy conclusion, ending with the collapse of the Iraqi regime as signified by the fall of Hussein's stronghold Tikrit on April 15, 2003. Later, in December 2003, the United States captured Hussein himself. The international community was divided on the legitimacy of this invasion, and the United States received heavy criticism from Belgium, Russia, France, China, Germany, and the Arab League.

The U.S. military strategy in Operation Iraqi Freedom was designed to be quick and surgically effective against designated targets, with minimum loss of life. The underlying assumption was that precision-guided munitions could wreak havoc unimpeded, aided by night missions, nullifying the capacity of the enemy to fight back. A key underlying component of this military strategy was dubbed "shock and awe," an attemp to reduce the enemy's will to fight through a display of overwhelming force through superior technology. Throughout the campaign, B-2 long-distance stealth bombers and F-117 stealth attack fighters evaded Iraqi radar and pounded key targets. Precision-guided munitions repeatedly hit military units, command-control-and-communications sites, and leadership hideouts.

Unfortunately, despite the speed of the war and casualty counts far lower than those in the Gulf War, significant collateral damage occurred:

> As the battles dragged on week after week, it became evident that the unrelenting mathematics of war had reached a tipping point. American and British forces had used such a high number of smart bombs that the normal miss rate of the Joint Direct Attack Munitions (JDAMs) and Tomahawk cruise missiles had produced an inordinate number of civilian casualties.
>
> The most ominous sign, clearly documented on Al-Jazeera, but not on U.S. news networks, was the daily growing number of women and children with missing limbs. . . . Doctors in Baghdad were so overwhelmed with casualties that they had neither the time, anesthetics nor antibiotics to repair complex injuries.[66]

While the exact toll is not known, on March 27, 2003, Iraqi health minister Omeed Medhat Mubarak said the total number of civilian injured and dead was more than 4,000, including 350 dead;[67] and the Iraq Body Count Database recently claimed that the number of civilians reported killed in Iraq by U.S.-led military action during the war was between 5,531 and 7,203.[68] Reportedly "only two out of every 10 smart bombs exploded where they should have"; even the U.S. Air Force admitted that out of 14,910 precision-guided munitions dropped in the first month of the war, 2,982 were expected to land off-target.[69] Aside from the interference provided by the worst dust storms in decades and key targets located in heavily populated areas, some precision-guided munitions evidenced guidance problems. Some Tomahawk cruise missiles missed their targets because of mechanical malfunctions, and on March 30 the U.S. military had to temporarily suspend Tomahawk launches from the eastern Mediterranean Sea and the Red Sea because some missiles had strayed into Turkey and Saudi Arabia.[70] So the increased sophistication of precision-guided munitions used in this conflict did not mean mishaps were nonexistent.

Case Reflections

A few patterns spanning these four cases appear noteworthy. Two of the cases (Allied Force and Enduring Freedom) involved more challenging subnational nonstate targets, while two (Desert Storm and Iraqi Freedom) involved a state's formal military forces as the target. Gathering intelligence and distinguishing between civilians and combatants appeared to be easier in Desert Storm than in either Allied Force or Enduring Freedom, given the more conventional nature of the warfare. So while the state of precision-guided weaponry was much more primitive in Desert Storm than in the later campaigns, the greater simplicity in targeting may have somewhat compensated for this deficiency; Iraqi Freedom, on the other hand, came closest to experiencing the best of both worlds. The special difficulties in identifying key vulnerability points in Allied Force compounded problems in that operation. Although it may be too soon to judge definitively, the apparently genuinely unified intense resolve of the ideologically motivated target terrorists in Afghanistan seems to have provided more of a military challenge than the possibly artificially grafted sense of regime loyalty present in Iraq or Serbia. Regardless, despite the relatively sophisticated countermeasures employed by the Serbs in Allied Force, none of the foes in these conflicts was able to disarm, destroy, or commandeer for their own purposes the precision-guided weaponry they faced. Although each case involved unquestioned military victory by the United States and its allies, which possessed overwhelming advantages in military force, and each achieved its immediate goal—freeing people in Kuwait, Serbia, Afghanistan, and Iraq from oppressive rule—none has yet unquestionably succeeded in achieving in target areas the long-term political changes of internal stability and the establishment of a democratic civil society adhering to global norms.

Dangers of Overreliance on Precision-Guided Munitions

Key dangers emerge if initiators depend too much on precision-guided munitions, where security policymakers utilize precision weaponry as the primary coercive instrument for all types of military uses abroad. If advocates of this technology succeed in promoting it as a panacea, some of these fears could well become realities.

First, the military itself may suffer from overdependence on precision-guided munitions. Despite the possibility of using precision warfare from air, land, and sea, too great a reliance on this approach could easily lead to the primacy of air power in future wars, a decidedly unwise course in cases where combat outcomes ultimately need to be determined by "boots on the ground." More generally, the self-conception of soldiers and the notion of military professionalism might very well change to focus more on managing and integrating weapons technologies rather than on developing and demonstrating personal courage and combat skills during wartime. From a societal perspective, this trend could also make the conduct of war more unintelligible to the civilian citizenry.

A second danger revolves around the unrealistic and sometimes foolish expectations developed by onlookers (including the domestic public and foreign allies) about the complete absence of unintended carnage generated by precision weaponry. For example, some conclude that, "if America's weapons are so superb, the accidental destruction of innocents must represent a contempt for human life and be attributable to willful negligence rather than the fog of war."[71] Although using dumb weapons clearly does not readily increase the plausible deniability for any accidents by the initiator, as, especially for the United States, blame will be assigned regardless, using precision weapons can equally clearly increase widespread international agreement on the initiator's culpability:

> Precision can have negative consequences even for casualty-sensitive nations. By vastly reducing the number of misses, precision leaves the United States open to greater criticism when a mistake happens. It is more difficult to explain away to an outraged foreign public an errant cruise missile than an errant "dumb bomb," since misses with the latter instrument are expected. Statements from Chinese officials suggest that this phenomenon fed beliefs within China that NATO's bombing of its Belgrade embassy in May 1999 was deliberate.[72]

Despite the tangible increases in efficiency provided by precision-guided munitions, even the smallest mistake by an initiator—due to human or mechanical error—can produce this kind of scorn.

As a result of the reduction in casualties and civilian damage associated with precision-guided munitions, treating them as a routine primary instrument of military coercion can eradicate initiators' prudent sense of political and diplomatic restraint in foreign military policy, leading to injudicious decisions about weighty missions. Precision-guided munitions may be especially appealing to political elite already predisposed to favor military action, as "the advent of precision weapons—and the ability to deliver those weapons with minimal risk to U.S. forces—has chipped away at old inhibitions regarding the use of force," and "the policy elite has become comfortable not just with the notion of possessing great military power, but of using it."[73] In this way, precision weaponry could promote a no-cost global dominance for the United States:

> High-tech military capabilities—to strike with accuracy and impunity, to anticipate and parry attack—reduce the uncertainty formerly inherent in the use of force. Technology largely obviates the need for sacrifice. It seemingly permits the United States to pursue global policies without subjecting the home front to the unwelcome dislocations of large-scale armed conflict—political protest or economic instability, for example. In short, technology enables the United States to use its military power to sustain American hegemony without the necessity of fighting messy old-fashioned wars.[74]

Ironically, wars or interventions using precision weaponry could signify to some not only a lack of resolve and willingness to take risks but also a lack of premeditated calculated caution about the true costs of war.

Finally, heavy reliance on precision technology could cause it, under certain limited circumstances, to diffuse inadvertently to those far less restrained than the initiating state. Given the porous flow of arms across national boundaries, one could easily imagine, for example, unscrupulous regimes using illegally obtained precision-guided munitions against their own people, or rogue states using them irresponsibly against innocent external targets. However, while small precision-guided munitions such as Stingers could easily fall into enemy hands, special obstacles reduce this concern for large-scale precision weaponry: it is more difficult for adversaries to utilize these arms since they tend to be high-tech, requiring extensive training and expertise, special maintenance infrastructures, detailed targeting intelligence, and in many cases large, expensive transport and launch platforms. Both components of large precision-guided munitions—the military command-and-control structure and the hardware and software technology embedded in weapons guidance systems—are difficult, but obviously not impossible, for an outsider to replicate due to the nature of the technologies, the difficulty of building them into a working system, and the operational challenges.

Conditional Utility of Precision Weaponry

A few highly tentative propositions emerge about the wartime utility of precision-guided munitions to help refine our understanding of these coercive instruments. Naturally, the inherent fallibility of all weapons and, specifically, precision-guided munitions' design and logistical limitations, prevent universal effectiveness in war. For example, "Simple map-reading errors in difficult terrain may well allow weapons to go astray. . . . [Q]uite apart from human error, the guidance systems of bombs may fail to work correctly or weapons may not come off the aircraft cleanly."[75] Although precision weapons increase the chances of destroying an identified target, "the problem of target acquisition remains," and topography and climatic conditions—along with electronic countermeasures—may interfere with their effectiveness.[76] Dispersion or simple decoys can easily thwart precision-guided munitions, and "concealment and camouflage may work very well" against them.[77]

Perhaps the most central limitation on the utility of precision-guided munitions is the high level of target intelligence required to exploit their capabilities. Pinpoint information is crucial. "However accurate the weaponry, an attacking force must know exactly what it is hoping to strike";[78] otherwise the target might be wrongly identified by intelligence.[79] For this reason it is generally agreed that "PGMs need more and better quality targeting information than more conventional (unguided) weapons":[80]

> Without intelligence to provide targets, PGMs are effectively useless. Dumb bombs still needed intelligence in order to be effective, but they are used en masse to destroy large targets, which were easier to find. PGMs, because of their precision, target much smaller, more individual items. Therefore the intelligence that is needed to select targets for PGMs needs to be that much more discerning. In the context of a

precision strike, targets have to be meticulously chosen and the choreography of a conflict becomes ever more essential.[81]

The tighter aim-points justify the more detailed intelligence needs of precision weaponry, to take advantage of the superior targeting capabilities of smart weapons and allow them to perform up to their technical specifications. While the ability to provide coordinates for mobile targets may be most critical,[82] remote surveillance systems cannot show where the leader of the country is sleeping at night or the location of a mobile command post.[83]

So while the need for timely and accurate intelligence is a military constant, precision-guided munitions thus appear to depend far more for wartime effectiveness on high-quality intelligence than does indiscriminate brute-force weaponry. Outside of direct targeting data, the most difficult and vital kind of intelligence to gather (affecting the decision about whether to use precision weaponry) may be intelligence that can "accurately predict how opponents will adjust to damage."[84] Despite recent advances in air and space-based intelligence, surveillance, and reconnaissance systems, "determining the effects of a strategic attack or an enemy's reaction to it remains an inexact science at best."[85]

Second, concern by initiators of wars or foreign interventions about civilian casualties and collateral damage can be a crucial discriminating condition affecting the utility of precision-guided munitions. On the one hand, the possession of precision-guided munitions could lead to demands among allies and enemies for absolutely minimal unintended damage,[86] especially if there is great external dependence on resources or international sympathy toward people within the target area. On the other hand, when large numbers of assaults occur over a long period of time, the result can be callous acceptance; among many possible examples, "the bombing of Afghanistan, which was at first such a shock to the international system, is rapidly becoming a bloody, botched but banal fact of life."[87] Often when one's target is a specific disruptive nonstate group within a country, the worry about civilian casualties and collateral damage is highest, for if an entire state is the enemy there is less concern with accidental destruction of an erroneous internal target.

Targets may try to take advantage of an initiator's concern about casualties and collateral damage. If existing political and physical restraints allow a targeted group to mix civilians and combatants together, even the most zealous advocates of pinpoint weaponry may find the accuracy insufficient to justify their use.[88] The result can sometimes be exactly the opposite of what an initiator wants from precision: "U.S. sensitivity *invites* adversary practices designed to put at risk the very civilians the United States seeks to leave unharmed," and "the typical U.S. responses to these practices then *reward* them in that the United States places further constraints on its own threats of force."[89] Such target mixing has been common in recent conflicts as a means of thwarting opposing forces' firepower.[90] Interspersing civilian and military targets seems especially possible in dense urban settings, making them generally poor choices for precision weaponry. "Fielded forces can reduce their vulnerability [to precision weaponry] by 'hugging' the civilian population or by taking cover in forests and urban areas";[91] in urban settings "targets constantly are on the move," while precision-guided munitions cannot be retargeted after launch;[92] and in cities "the fog of war remains to a considerable degree impenetrable even to the latest technology" at least in part because "the technology that aims to distinguish between combatants and non-combatants is very much at the wish list stage."[93] Furthermore, "nonstate actors are often particularly agile at exploiting human shields and blurring combatant-noncombatant distinctions,"[94] with insurgents or terrorists especially hard to isolate. Thus, during armed conflict, precision-guided munitions appear to be more effective than indiscriminate brute-force weaponry when an initiator cares about civilian casualties and collateral damage and when mixing civilian and combatant targets together is not feasible.

A third condition relates to the passion of the target. During wartime, precision-guided munitions appear to be less effective than indiscriminate brute-force weaponry if a target possesses large advantages in its enthusiastic resolve for its cause (asymmetric will). This ratio of resolve usually boils down to the determination of each side to keep on fighting "until the last man" and their relative willingness to sacrifice and suffer in the process.[95] This resolve is especially significant if a target does not properly extrapolate current damage from precision weaponry to more widespread and devastating damage in the future. For reasons related to both cognitive and emotional limitations, adversaries may overestimate their own resilience in the face of significant losses, or alternatively underestimate the initiator's will to continue the attack. Following this logic, a truly irrational target with passionate resolve for its cause may leave little alternative to total obliteration by an initiator utilizing dumb maximally destructive weaponry. Most important, adversaries may interpret heavy reliance on precision-guided munitions as signaling a lack of will on the part of the initiator, weakening deterrence prospects:

> Indeed, the more the United States relies on high-technology solutions to address political anxieties about casualties and collateral damage, the more adversaries have reason to doubt U.S. credibility. They may conclude, for example—as did Mohammed Farah Aideed, Saddam Husayn, and others before them—that the U.S. emphasis on minimizing casualties means that the United States will back down in the face of combat deaths.[96]

Although foreign perceptions of U.S. military technology tend to be favorable, with its superiority universally acknowledged, the sustained U.S. will is in question to bring this power to bear over the long haul against a particular adversary in a particular conflict.

When facing an adversary possessing this staunch conviction and advantage in resolve, the wartime challenge is lower if the leadership alone is passionate for a cause. Precision weaponry appears more useful when an initiator thinks that the bulk of an enemy's population can be rehabilitated (for example, if it thinks that democracy can eventually be put in place, or that not all of the society is evil or corrupt) than when it deems the whole society to be a den of vipers. If the leadership alone is a target, particularly if all the leaders are gathered in one location (such as a small facility in the desert), then it may be feasible to use precision-guided munitions as an element in a classic regime-decapitation strategy.

Finally, precision-guided munitions appear more effective than indiscriminate brute-force weapons in armed conflicts when there are specific identifiable enemy vulnerability points, which if destroyed can cripple a foe's capacity to fight. The logic behind precision attack here is that it is possible to foster rapid enemy capitulation by knocking out a few nodes critical to the economy or the armaments industry. Although dumb bombs can be used against such targets, they seem much less efficient because an initiator would require a larger number of sorties to achieve the desired damage. Since World War II, the United States has been examining complex target systems to identify the key critical nodes, elimination of which would be sufficient to bring down the enemy's whole system. For example, a ball bearing plant may be the choke point for manufacturing military systems, or a structure holding up a single tank in an oil refinery may be absolutely vital to continued oil production. Although "wholesale attacks on population centers do little to break the enemy's will to resist," in many cases "the surgical removal of an enemy's most vital elements should make it easier for him to surrender."⁹⁷ And, after conflicts end, this approach could make restoration of the attacked system faster, easier, and cheaper.

Conversely, when such critical nodes do not exist in enemy territory, dumb weaponry may fit the bill exactly. Some targets are ideally suited to dumb iron bombs, including large, soft targets such as rail yards, industrial complexes, tank farms, and deployed ground formations of troops.⁹⁸ To be successful in such attacks, one just needs a large amount of explosive power delivered in the approximate area of the target. A dozen B-52 Stratofortress intercontinental bombers dropping tons of iron bombs, obliterating everything in sight, are a maximally destructive but relatively imprecise ordnance that can sometimes be more terrifying—and serve better to prevent future violent incidents—to both observers and victims than precision-guided munitions.

Conclusion

As with all armaments, there is a time and a place for the use of precision weaponry. This chapter's analysis points to a few clusters of conditions—including high target intelligence, initiator concern about human life, lack of passionate resolve on the part of the target for its cause, and identifiable target vulnerability points—that enhance the wartime utility of precision-guided munitions (with the reverse circumstances highlighting the utility of indiscriminate brute-force weaponry). One irony is that exactly when precision weaponry seems on the surface to be most needed, such as in fighting nonstate terrorist or insurgent groups in urban settings, its effectiveness may end up being lowest because some targeting discriminations are not simply a function of advanced technology.

Without due consideration to precision-guided munitions' wartime utility limits, the allure of advanced precise military technology could eventually lead to realization of the dangers of inadvertent diffusion, erosion of the military image, unrealistic expectations about the absence of carnage, elimination of a prudent sense of diplomatic restraint, and emergence of unending "bloodless" conflicts. However, a broad examination of history reassures us that human strategy, not military technology, has traditionally determined the course of armed conflict: "Until the present time, the application of sound imaginative thinking to the problems of warfare (on either an individual or institutional basis) has been more significant than any new weapon."[99] Nonetheless, we cannot rest secure by letting this comforting pattern relieve us for one moment from the responsibility for seriously scrutinizing the relative utility of the tools of combat we create.

Notes

1. This chapter draws heavily from Robert Mandel, "The Wartime Utility of Precision Versus Brute Force in Weaponry," *Armed Forces & Society* 30 (winter 2003): 171–200.

2. John J. Mearsheimer, "Precision-Guided Munitions and Conventional Deterrence," *Survival* 21 (March–April, 1979): 68; and Stephen Biddle, "Land Warfare: Theory and Practice," in John Baylis, James Wirtz, Eliot Cohen, and Colin S. Gray, eds., *Strategy in the Contemporary World* (New York: Oxford University Press, 2002), p. 103.

3. Daniel Goure and Gordon McCormick, "PGM: No Panacea," *Survival* 22 (January–February, 1980): 15–17; and Michael Schrage, "Too Smart for Our Own Good," *Washington Post*, June 2, 2002, p. B3.

4. Mark Thompson, "The Lessons of Afghanistan," *Time* 159 (February 18, 2002): 29.

5. James Digby, *Precision-Guided Munitions*, Adelphi Paper no. 118 (London: International Institute for Strategic Studies, summer 1975), p. 1.

6. Stephen L. McFarland, *America's Pursuit of Precision Bombing, 1910–1945* (Washington, DC: Smithsonian Institution Press, 1995), p. ix.

7. Digby, *Precision-Guided Munitions*, p. 1.

8. Robert A. Pape, *Bombing to Win: Air Power and Coercion in War* (Ithaca: Cornell University Press, 1996), pp. 4–5.

9. Michael Mok, "Precision Guided Munitions" (2002), available online at www.georgetown.edu/sfs/programs/stia/students/vol.03/mok_pgm.htm.

10. Eliot A. Cohen, "A Revolution in Warfare," *Foreign Affairs* 75 (March–April 1996): 45.

11. Max Boot, "The New American Way of War," *Foreign Affairs* 82 (July–August 2003): 53.

12. Robert E. Osgood, "The Expansion of Force," in Robert J. Art and Kenneth N. Waltz, eds., *The Use of Force: International Politics and Foreign Policy* (Boston: Little, Brown, 1971), p. 50.

13. Bruce Russett and Harvey Starr, *World Politics: The Menu for Choice*, 5th ed. (New York: W. H. Freeman, 1996), p. 146.

14. Gil Merom, *How Democracies Lose Small Wars: State, Society, and the Failures of France in Algeria, Israel in Lebanon, and the United States in Vietnam* (New York: Cambridge University Press, 2003), pp. 36–37.

15. Lawrence Freedman, ed., *Strategic Coercion* (New York: Oxford University Press, 1998), p. 23.

16. Sir Basil Henry Liddell Hart, *Thoughts on War* (London: Faber and Faber, 1944), p. 42; and Lieutenant-General Buster C. Glosson, "Impact of Precision Weapons on Air Combat Operations," *Airpower Journal* (summer 1993), available online at www.airpower.maxwell.af.mil/airchronicles/apj/glosson.html.

17. Daniel Byman and Matthew Waxman, *The Dynamics of Coercion: American Foreign Policy and the Limits of Military Might* (New York: Cambridge University Press, 2002), pp. 4–5.

18. Thomas C. Schelling, *Arms and Influence* (New Haven: Yale University Press, 1966), p. 3.

19. Byman and Waxman, *Dynamics of Coercion*, p. 5.

20. Steven Metz, "Strategic Asymmetry," *Military Review* 81 (July–August 2001): 30.

21. Paul G. Gillespie, "Ultimate Weapons: The Development of Precision Guided Munitions and Their Effect on National Security Policy," unpublished paper.

22. Byman and Waxman, *Dynamics of Coercion*, p. 232.

23. Digby, *Precision-Guided Munitions*, p. 1.

24. John D. Gresham, "A New Generation Emerges: Precision Guided Munitions" (1999), available online at www.aviation100.com/web04/yid/precision.html.

25. Robert K. Ackerman, "Afghanistan Is Only the Tip of the Network-Central Iceberg," *Signal Magazine* (April 2002), available online at www.us.net/signal/currentissue/april02/afghanistan-april.html.

26. Byman and Waxman, *Dynamics of Coercion*, p. 232.

27. Pape, *Bombing to Win*, pp. 321–323.

28. Gresham, "New Generation Emerges."

29. Byman and Waxman, *Dynamics of Coercion*, pp. 231–232.

30. Phillip S. Meilinger, "A Matter of Precision: Why Air Power May Be More Humane Than Sanctions," *Foreign Policy* 123 (March–April 2001): 78–79.

31. Richard P. Hallion, "Precision Guided Munitions and the New Era of Warfare," Australian Air Power Studies Centre Paper no. 53 (1995), available online at www.fas.org/man/dod-101/sys/smart/docs/paper53.htm.

32. Meilinger, "Matter of Precision," pp. 78–79.

33. J. F. C. Fuller, *Armament and History: The Influence of Armament on History from the Dawn of Classical Warfare to the End of the Second World War* (New York: Da Capo Press, 1998), p. 20.

34. Colonel Trevor N. Dupuy, *The Evolution of Weapons and Warfare* (Indianapolis, IN: Bobbs-Merrill, 1980): pp. 286–287.

35. Digby, *Precision-Guided Munitions*, p. 1.

36. Mok, "Precision Guided Munitions."

37. Harvey M. Sapolsky and Jeremy Shapiro, "Casualties, Technology, and America's Future Wars," *Parameters* (summer 1996): 121.

38. George Friedman and Meredith Friedman, *The Future of War: Power, Technology, and American World Dominance in the Twenty-First Century* (New York: St. Martin's Griffin, 1996), p. 35.

39. Thompson, "Lessons of Afghanistan," p. 32.

40. Pape, *Bombing to Win*, pp. 319–320.
41. Cohen, "Revolution in Warfare," p. 40.
42. Osgood, "Expansion of Force," pp. 49–50.
43. Byman and Waxman, *Dynamics of Coercion*, p. 229.
44. Michael E. O'Hanlon, "A Flawed Masterpiece," *Foreign Affairs* 81 (May–June 2002): 52.
45. Jennifer Pangyanszki, "Lessons Learned from New-Era Warfare," *CNN Report* (April 19, 2003), available online at www.cnn.com/2003/us/04/19/sprj.irq.lessons.
46. Anthony H. Cordesman, "The 'Instant Lessons' of the Iraq War: Main Report" (Washington, DC: Center for Strategic and International Studies, May 14, 2003), p. 143.
47. U.S. General Accounting Office, *Operation Desert Storm: Evaluation of the Air Campaign* (Washington, DC: U.S. Government Printing Office, June 1997), p. 31.
48. CNN News, "Hit Smarter, Not Harder? Gulf War Strikes Marked a Sea Change in Air Tactics" (2001), available online at www.cnn.com/specials/2001/gulf.war/legacy/airstrikes.
49. Cohen, "Revolution in Warfare," pp. 39–40.
50. Mark J. Conversino, "The Changed Nature of Strategic Air Attack," *Parameters* 27 (winter 1997–1998): 28–41.
51. Andrew J. Bacevich and Eliot A. Cohen, "Strange Little War," in Andrew J. Bacevich and Eliot A. Cohen, eds., *War over Kosovo* (New York: Columbia University Press, 2001), p. ix.
52. David A. Ochmanek, "What Operation Allied Force Can Teach Us About Capabilities Needed for Future Military Operations" (1999), available online at www.ndu.edu/inss/symposia/topical99/ochmanekpaper.html.
53. William M. Arkin, "Operation Allied Force: 'The Most Precise Application of Air Power in History,'" in Bacevich and Cohen, *War over Kosovo*, pp. 21–22.
54. Anatol Lieven, "Hubris and Nemesis: Kosovo and the Pattern of Western Military Ascendancy and Defeat," in Bacevich and Cohen, *War over Kosovo*, p. 100.
55. Earl H. Tilford Jr., "Operation Allied Force and the Role of Air Power," *Parameters* (winter 1999–2000): 25.
56. Lieven, "Hubris and Nemesis," p. 120.
57. Arkin, "Operation Allied Force," pp. 14–15.
58. Schrage, "Too Smart for Our Own Good," p. B3.
59. Michael Kelly, "The American Way of War," *Atlantic Monthly* 286 (June 2002): 16.
60. "Operation Enduring Freedom," unpublished article (March 16, 2004), available online at www.globalsecurity.org/military/ops/enduring-freedom-ops.htm.
61. Kelly, "American Way of War," p. 16.
62. Sandra I. Erwin, "Air Warfare Tactics Refined in Afghanistan," *National Defense Magazine* (April 2002), available online at www.nationaldefensemagazine.org/article.cfm?id=75.

63. Carl Conetta, "Operation Enduring Freedom: Why a Higher Rate of Civilian Bombing Casualties?" (Cambridge, MA: Commonwealth Institute, Project on Defense Alternatives, January 18, 2002; rev. January 24, 2002), available online at www.comw.org/pda/0201oef.html.

64. Thomas E. Ricks, "Bull's-Eye War: Pinpoint Bombing Shifts Role of GI Joe," *Washington Post,* December 2, 2001, p. A1.

65. Fred Kaplan, "New Warfare: High-Tech US Arsenal Proves Its Worth," *Boston Globe,* December 9, 2001, p. A34.

66. Jim Wilson, "Smart Weapons Under Fire," *Popular Mechanics* 180 (July 2003): 42–43.

67. Fox News, "Iraq Accuses US of Targeting Civilians" (March 27, 2003), available online at www.foxnews.com/story/0,2933,82309,00.html.

68. "Iraqi Body Count" (2003), available online at www.iraqbodycount.net/bodycount.htm.

69. Wilson, "Smart Weapons Under Fire," p. 43.

70. Fox News, "Six Thousand Precision-Guided Bombs Dropped on Iraq Since War Began" (March 29, 2003), available online at www.foxnews.com/story/0,2933,82559,00.html.

71. Schrage, "Too Smart for Our Own Good," p. B3.

72. Sheila Melvin, "Why Chinese Can Believe Worst About U.S. Bombing," *USA Today,* May 12, 1999, p. 15A; Elisabeth Rosenthal, "Public Anger Against U.S. Still Simmers in Beijing," *New York Times,* May 17, 1999, p. A1; Eliot A. Cohen, "Kosovo and the New American Way of War," in Bacevich and Cohen, *War Over Kosovo,* p. 55; and Byman and Waxman, *Dynamics of Coercion,* p. 232.

73. Ricks, "Bull's-Eye War," p. A1.

74. Andrew J. Bacevich, "Neglected Trinity: Kosovo and the Crisis in U.S. Civil-Military Relations," in Bacevich and Cohen, *War over Kosovo,* p. 180.

75. Jonathan Marcus, "Why 'Precision Bombing' Goes Off Course," *BBC Online Network* (June 1, 1999), available at http://news.bbc.co.uk/hi/english/world/europe/newsid_358000/358153.stm.

76. Goure and McCormick, "PGM," p. 17.

77. Ken Silverstein, "Buck Rogers Rides Again: A 'Revolution' in High-Tech Systems Promises Big Profits and for the U.S. Risk-Free War," *Nation* 269 (October 25, 1999): 23; and Digby, *Precision-Guided Munitions,* p. 7.

78. Jonathan Marcus, "Smart Weapons in Forward Role," *BBC Online Network* (February 20, 1998), available at http://news.bbc.co.uk/hi/english/special_report/iraq/newsid_53000/53837.stm.

79. John Nichol, "The Myth of Precision: However Sophisticated the Technology, Bombing Campaigns Will Always Kill Innocent People," *The Guardian* (October 29, 2001), available online at www.bard.edu/hrp/hrresponses/terrorism/nichol.10.29.01.htm.

80. Gresham, "New Generation Emerges."

81. Mok, "Precision Guided Munitions"; and William Arkin, "Smart Bombs, Dumb Targeting?" *Bulletin of the Atomic Scientists* 56 (May–June 2000): 50.

82. Kaplan, "New Warfare," p. A34.

83. Silverstein, "Buck Rogers Rides Again," p. 23.
84. Pape, *Bombing to Win,* p. 321.
85. Conversino, "Changed Nature of Strategic Air Attack," p. 38.
86. Schrage, "Too Smart for Our Own Good," p. B3.
87. Gary Younge, "Peace by Precision," *The Guardian* (October 29, 2001), available online at www.guardian.co.uk/comment/story/0,3604,582444,00.html.
88. Schrage, "Too Smart for Our Own Good," p. B3.
89. Byman and Waxman, *Dynamics of Coercion,* pp. 146–147.
90. See, for example, Schelling, *Arms and Influence,* p. 4; and Barbara Crossette, "Civilians Will Be in Harm's Way If Baghdad Is Hit," *New York Times,* January 28, 1998, p. A6.
91. Cohen, "Kosovo," p. 55.
92. Sandra I. Erwin, "Air Force Wants Missiles Redirected in Flight," *National Defense Magazine* (May 2003), available online at www.national defensemagazine.org/article.cfm?id=1101.
93. Lieven, "Hubris and Nemesis," pp. 113–114.
94. Byman and Waxman, *Dynamics of Coercion,* p. 196.
95. Metz, "Strategic Asymmetry," p. 27.
96. Byman and Waxman, *Dynamics of Coercion,* p. 232.
97. Glosson, "Impact of Precision Weapons."
98. Hallion, "Precision Guided Munitions."
99. Dupuy, *Evolution of Weapons and Warfare,* p. 340.

4

Nonlethal Weaponry

Compared to the other two instruments of the quest for bloodless war—precision-guided munitions and information warfare—nonlethal weaponry is still considered rather unusual and has received far less public discussion. Designed for application in a narrower range of coercive confrontations, the uses of these less-than-deadly arms are considerably less flashy, though not inherently less effective. Being the least tried-and-true of the three approaches, without a lot of testing during recent major international conflicts, the potential application of nonlethal technologies in warfare as a means of casualty aversion is somewhat wrapped in mystery.

Nonlethal weaponry is increasingly becoming available for widespread application as a means of promoting national security.[1] Spurred on by technological advancements permitting forms of force not even conceived of in prior eras, this new instrument raises many fascinating and important questions about the effective ways of fighting and managing conflict in today's world. Among these unresolved issues is the extent to which nonlethal weapons can advance the cause of casualty aversion on a global level: since nonlethal weaponry has been used more in nonwar situations within states, employing it in violent international conflicts is still something of a novelty. From lofty concerns about morality to practical worries about battlefield backfire effects, there appears to be a significant gap in understanding. Fundamental underlying issues emerge about the nature of weaponry, the military, and even warfare itself, with symbolic psychological implications being as important as tangible field impact.

Even in the aftermath of the terrorist attacks of September 11, 2001, the nonlethal weaponry issue emerged front and center. In particular, a debate raged about whether pilots on commercial airlines ought to carry nonlethal weapons in the form of stun guns in case unruly passengers attempted to take over the aircraft, with some arguing that the presence of such weapons would increase danger to innocent bystanders, and others contending that they would be almost completely ineffective against terrorists. The disagreements that occurred about the advantages and disadvantages of this potentially new form of transportation security served to highlight the general ignorance about when nonlethal weaponry is most and least effective.

After a background discussion of definition, motives, history, and dangers, this chapter specifically develops some highly tentative propositions about the conditions under which nonlethal weaponry is most beneficial—and by implication detrimental—as an instrument of national and international security. Through this analysis it is possible to see how policy makers would have greater awareness when to advocate or oppose its use as a coercive casualty-minimizing instrument for managing foreign turmoil.

Existing literature on nonlethal weaponry—outside of largely classified within-government documents—is unfortunately extremely thin, with the conceptual context largely unexplored. Those writings available on this topic largely contain sweeping argumentation either supporting or opposing the use of this kind of armament in future combat situations. Moreover, the relative paucity of major examples of publicly documented recent nonlethal weaponry applications makes it difficult to draw empirically based assessments of the current and projected utility of this coercive instrument. While these obstacles do not eliminate the possibility of intelligent discussion of nonlethal weaponry, they do suggest that any such discussion would be inescapably highly preliminary in its conclusions.

Definition of Nonlethal Weaponry

The controversies swirling around nonlethal weaponry begin with the substantial difficulties involved in delineating this security instrument. The U.S. Defense Department provides a very useful starting point, defining nonlethal weaponry as "weapons that are explicitly designed to and primarily employed so as to incapacitate personnel and materiel, while minimizing fatalities, permanent injury to personnel, and undesired damage to property and the environment," where "unlike conventional lethal weapons that destroy their targets principally through blast, penetration, and fragmentation, non-lethal weapons employ means other than gross physical destruction to prevent the target from functioning."[2] According to this definition, the identity of a weapon as nonlethal thus must be inherent in its design, not just in its mode of employment. There is explicitly no underlying assumption that deaths are eliminated, just that they are minimized; and no minimum quantitative threshold for lethality is identified (such as no more than 10 percent of the target population being severely affected) so as to maximize flexibility in design. Unlike other nontraditional arms, such as chemical or biological agents, nonlethal technologies can actually reduce the scope of devastation and destruction.

Several definitional problems are readily apparent. Is a small explosive charge designed for contained demolition of structures and detonated far from known human populations a nonlethal weapon? Is it proper to classify a foam barrier as a weapon at all? How can one discuss the level of damage beneath which a weapon would properly be classified as nonlethal? The context may be critical in determining nonlethality: "Biological or chemical agents that destroy crops without directly affecting people would still be considered lethal if starvation is the likely result; a microwave weapon that disables a truck that subsequently drives off a cliff, killing the driver, would be nonlethal [while] the same weapon used against a helicopter in flight would have to be considered lethal."[3] The reality is that such a wide variety of coercive instruments fall under this rubric that it is nearly impossible to develop a tight delineation that completely and unambiguously indicated how to distinguish nonlethal from lethal weaponry. Nonetheless, despite this inherent fuzziness, it is possible at least to get a general handle on what is included and excluded.

While there are many ways to subdivide types of nonlethal security instruments, reviewing a few common typologies seems useful. A dominant categorization scheme used by the military distinguishes among acoustic, biotechnical, chemical, electromagnetic, mechanical, and optical forms of nonlethal technology.[4] The range of specific nonlethal security instruments is broad and constantly evolving, including such coercive techniques as blunt projectiles, tear gas, traction modifiers, nets or rapid-hardening rigid foam, radio frequency or microwave technologies, noxious smells, and acoustical interference. A more colloquial enumeration is "slickums, stickums, super acids, goop guns, blinding lasers, non-nuclear electromagnetic pulses, high power microwaves, laser weapons, infrasound, computer viruses, and metal-eating microbes."[5] It is possible to split nonlethal security instruments according to whether they primarily serve disabling access denial functions (preventing entrance to a zone) or enabling combat support functions. Alternatively, one could divide nonlethal technologies between counterpersonnel measures, involving clearing facilities and structures of personnel, incapacitating individuals, crowd control, and denying an area to personnel; and countermatériel measures, involving disabling equipment and facilities and denying an area to vehicles, vessels, and aircraft.[6] Figure 4.1 displays how these categorization systems could discriminate among different types of nonlethal instruments.

Motives Behind Nonlethal Weaponry

Nonlethal weaponry appears to be potentially relevant for several different kinds of broad security purposes under the general umbrella of combat support. Given the unorthodox nature of this security instrument, those seeking to employ it generally have had to specify quite explicitly the goals they have in mind. Recent increases in vulnerability of civilians, sensitivity to casualties, distortion of foreign policy

Figure 4.1 Classification of Nonlethal Weaponry

Defensive disabling access denial functions	*Counterpersonnel nonlethal measures*
Traction	Blunt or soft projectiles
Goop ejectors	Stinger grenades
Trapping nets	Stun guns
Rapid-hardening rigid foam	Tear gas
	Pepper spray
Offensive enabling combat support functions	Noxious smells
Blunt of soft projectiles	Blinding laser weapons
Stinger grenades	Acoustical interference
Super acids	
Metal-eating microbes	*Countermatériel nonlethal measures*
	Traction modifiers
Both	Electromagnetic pulses
Tear gas	Super acids
Pepper spray	Metal-eating microbes
Blinding laser weapons	
Stun guns	*Both*
Noxious smells	Goop guns
Acoustical interference	Trapping nets
Microwave technologies	Rapid-hardening rigid foam
Electromagnetic pulses	Microwave technologies

intent, negative media exposure, and narrowed field combat options have combined with a pattern of security failures to help identify the objectives of developing and possessing nonlethal weaponry.

Perhaps the noblest aim is a humane one, attempting to reduce death and severe injury among noncombatants, who have inadvertently become more common victims of warfare due to the increased destructive power of many modern lethal weapons technologies:

> Killing, even remotely or robotically, is what we want to avoid as much as possible. From this realization springs the growing interest in nonlethal weapons. Dozens of goos, sprays, traps, and noisemakers are being developed to disable enemy equipment and personnel. In this kit, we hope, is (or will be soon) just the alternative we need for those times when a lethal encounter is undesirable.[7]

Some analysts even make the grandiose argument that nonlethal weaponry is "a revolutionary catalyst that will alter the very nature of war fighting towards a more benign arena in the twenty-first century."[8] After all, some nonlethal technologies are so incapacitating that they deny targets the physical possibility of achieving their objectives by removing the means to act. The military readily and pragmatically acknowledges, in the context of nonlethal weapons, that "any conflict has associated political objectives, and the accomplishment of those objectives is often enhanced by minimizing non-combat casualties and collateral damage."[9] Even when dealing with military personnel, equipment, and structures, the effects of nonlethal weaponry promise to be generally much more temporary and reversible than those of lethal weaponry. A side benefit of this nonpermanence is that the use of nonlethal weaponry could lower postconflict economic costs for restoring stability due to the reversibility of the impact in target societies.

A related motive deals with improving internal and external public image. With respect to the internal image, the attempt here is to attain higher levels of domestic social acceptability through military engagement involving lower levels of violence. Within the United States, for example, a perception persists that "the American public will no longer accept conflicts that result in high numbers of civilian casualties, and as public support is deemed crucial for any overseas deployment of US forces, non-lethal technologies are seen as vital to maintaining favorable public opinion."[10] More specifically, "in 'operations other than war,' the killing of civilians could call the continuation of the mission into question; for this reason, disabling [nonlethal] weapons are important due to the fact that they are specifically 'media-friendly' weapons."[11] Regarding external image, the effort is to minimize the undesired twisting of foreign policy military action for propaganda purposes:

> Non-lethal weapons provide decision-makers a new means to resolve difficult political situations. In peacetime operations, for example, significant work to conduct humanitarian assistance could be overshadowed if because of circumstances deadly force must be applied. Organizations hostile to U.S. assistance/intervention could exploit such an event through the media by presenting it as unnecessary violence and damage. Non-lethal technologies provide a means for precluding such deadly confrontations and denying the opportunity to exploit them for propaganda purposes.[12]

The use of nonlethal technologies also may reduce the chances for the creation of martyrs, around whom disruptive groups can rally. Sometimes, exposing casualties to the public eye occurs even when not premeditated with hostile intent:

> When such noncombatant casualties occur—even as the unavoidable result of actions taken under clear military necessity—they are immediately and graphically reported worldwide by networked media organizations. Such reporting often creates considerable local, international, or domestic U.S. opposition to the continued presence of U.S. forces in the area of crisis. This can result in the loss of perceived legitimacy and severely limit the utility of military force as a policy option in the furtherance of national interests. Clever opponents are quick to recognize these constraints and will seek to turn the situation to their own advantage.[13]

This reinterpretation of positively motivated great power actions is a major concern to these states, and nonlethal weaponry has the potential to circumvent this predicament.

A third, related goal motivating the development of nonlethal weaponry addresses military effectiveness, in which there is an effort to reduce the widely reported hesitation of soldiers to perform their duty on the battlefield under the constant glare of media coverage.[14] The military has come to believe that "there is public concern about the violent nature of modern warfare and about long, lethal, and costly campaigns where vital interests of the nation may not be clearly defined," particularly when dealing with operations "reported by the news media who can disseminate video observations, globally, in real time, creating an unprecedented level of scrutiny."[15] It has become increasingly difficult to isolate or inoculate military personnel from this kind of nonrequested transparency, which can serve indirectly to reduce their trained, automated responses to imminent danger.

Fourth, the prevalence of murky international confrontations, in which it is hard to isolate target groups and, thus, in which attacks may accidentally hit innocent victims, stimulates policymakers to consider nonlethal weaponry. These are chaotic situations in which, traditionally, lethal force has had problems:

> Some military operations other than war scenarios include the presence of paramilitary forces or armed factions which present a real but ill-defined threat. In these situations, the mission of military forces commonly has aspects that are *preventive* in nature. That is, military forces accomplish their mission by preventing individuals or groups from carrying on undesirable activities such as rioting and looting or attacking, harassing, and otherwise threatening opponents. Sometimes, hostile elements blend in with the local population of uninvolved citizens. Other times, sectors of the local population may rise against our forces and become active participants in acts of violence. Factional alignments, the level of violence, and the threat to mission accomplishment may change frequently and with little or no warning. Under such circumstances, the identity of our opponents is uncertain, and the use of deadly force for purposes other than self-defense may be constrained by rules of engagement or by the judgment of the commander on the scene.[16]

Enemies of the West have been known to purposely mix combatants with noncombatants because these foes realize the West's reluctance to take action that could harm innocent people. Transnational and subnational groups appear to pose greater problems than state targets in terms of unambiguous demarcation and accidental harm to innocent bystanders. However, it is important to recognize that, although the use of nonlethal weaponry in these scenarios definitely decreases the costs of coercive engagement—reducing the risk of killing innocent victims—there is no guarantee at all that it will increase the benefits of such engagement—achieving political-military objectives.

Although this volume's focus is on wartime casualty aversion, it is worth noting that a fifth, narrower purpose of nonlethal weaponry is to improve the effectiveness of peacekeeping operations. "Many advocates of non-lethal weapons point to the growth of peacekeeping and peace enforcement operations where new military force structures are evidently needed but where, at present, effective alternative non-lethal weapons systems are not available."¹⁷ In some ways, nonlethal weaponry seems very much in tune with the aims of peacekeeping, where the consent of those involved is critical. Despite nonlethal weaponry's general promise, their application in this area involves many unexplored complexities.

Ultimately, it must not be forgotten that one of nonlethal weaponry's most basic aims is to expand the range of choice for military commanders on the battlefield:

> Traditional military weapons require commanders to make difficult "trade off" decisions regarding the proper balance between mission accomplishment, force protection, and the safety of noncombatants. We may relax the rules of engagement in order to enhance mission accomplishment or force protection through increased freedom in the application of firepower, but this potentially decreases the safety of noncombatants. Conversely, when we increase the safety of noncombatants through restrictions on the use of firepower, our troops become potentially more vulnerable and their mission more difficult to achieve.
>
> Non-lethal weapons expand the number of options available to commanders confronting situations in which the use of deadly force poses problems. They provide flexibility by allowing U.S. forces to apply measured military force with reduced risk of serious noncombatant casualties, but still in such a manner as to provide force protection and effect compliance. Because we can employ non-lethal weapons at a lower threshold of danger, commanders can respond to an evolving threat situation more rapidly.¹⁸

It is commonly held that "one of the main purposes for having nonlethal weapons in [the U.S.] arsenal is to provide commanders and policy-makers additional options between no use of military force at all and the use of lethal military force; because this third option can provide a more humane, discriminate, and relatively reversible means of employing military force, with more precisely tailored and focused effects, they may be more appropriate for some missions than lethal weapons" in playing a role "across the spectrum of conflict, from low-intensity conflict through major regional engagements."¹⁹ With so much of the legacy weaponry in the West owing to the Warsaw Pact, in its design to help defeat an invasion of Western Europe, this widening of options appears to be extremely valuable.

This review of motivations indicates that a significant gap exists in security protection that nonlethal weaponry might be able to fill. Traditionally, security policymakers escalate from diplomacy to economic sanctions to military intervention, and nonlethal weaponry may present something of a "middle option."[20] Internationally, the capacity for generating widespread destruction is "gravitating into increasingly less responsible hands,"[21] including violent subnational and transnational insurgency and terrorist groups. For the first time since the emergence of the nation-state, more military weapons are in the hands of private citizens than in the hands of national governments due to the uncontrolled spread of arms.[22] Embarrassments such as the failed U.S. intervention in Somalia in 1992–1993, along with the chaos at the World Trade Organization meetings in Seattle, Washington, in winter 1999 and the Group of Eight summit in Genoa, Italy, in summer 2001, demonstrate the need for new approaches to preventing and managing internal and external turmoil.

History of Nonlethal Weaponry

Although never dominant or widely publicized, the use of nonlethal weaponry extends back to the very beginnings of armed warfare. Indeed, a common reference point for early uses of this coercive instrument has been the biblical story of Joshua in the Battle of Jericho, in which loud acoustical interference caused the protective walls to crumble and the campaign to be successful. It is easy to forget just how long this alternative coercive instrument has been around:

> It is important to point out that non-lethals are not new. They have existed for centuries and have supported many past operations. Smoke has long been used for concealment. Entangling devices, such as cal trops, were employed against cavalry and infantry long before they were used against motorized vehicles. So, while the term "non-lethals" may be relatively new, the military's experience with them is not.[23]

Always designed for special-purpose uses, nonlethal weaponry has served an important yet subsidiary role for ages.

Looking over the long history of military warfare, nonlethal weaponry was of course but one of several elements contributing to a reduction in violent carnage. Outside of the role of precision-guided munitions and information warfare, the presence of technological shielding and the absence of human commitment played major roles: "In the past, nonlethal warfare did not rely on the use of nonlethal weapons; rather, it was the fortuitous result of the superiority of body armor over offensive weaponry or the mutual lackadaisical approach of opposing soldiers and leaders."[24] Many new forms of nonlethal weapons exist today that did not exist in the past, and the post–Cold War setting has afforded different occasions for their use, but the contribution to casualty aversion continues.

Recent renewed interest in nonlethal weapons goes back at least a number of decades. For example, a report commissioned by the National Science Foundation in 1971 on possible uses for law enforcement called for the development of nonlethal weapons such as soft plastic ricochet rounds, tasers, and foam generators, and for many years immobilizing foams have served to safeguard nuclear facilities.[25] However, the major thrust of exploring this unorthodox form of security instrument was to wait until after the Cold War.

The distinctive set of dangers embodied by the post–Cold War security environment has triggered special attention by the great powers in this regard. In particular, exploration of nonlethal weaponry has resulted from the proliferation of different forms of anarchic conflict, including internal ethnic strife, terrorist and insurgent challenges to the state, transnational criminal organization activity, mob violence, urban warfare, and anarchist-triggered chaos. As the U.S. Defense Department itself points out, turning to nonlethal weaponry at least in part results from substantial changes in the nature of warfare:

> Since 1990, the role of U.S. military forces in military operations other than war (MOOTW) and military operations in urban terrain (MOUT) has increased dramatically. These nontraditional military operations (Bangladesh, Haiti, Somalia, Bosnia) demand greater flexibility because they commonly involve close and continual interaction between U.S. forces and noncombatant civilians. Operational use of non-lethal weapons (NLW) is driven by the increasing urbanization of warfare and recognition of the potential for massive collateral damage (unintended civilian casualties and property damage) caused by modern weapons of war.[26]

Nothing could be further from the traditional confrontation of opposing armies on an open battlefield far from populated areas.

The decreasing ability to isolate combat strictly in designated battle zones during wartime provides an opportunity for nonlethal weaponry to play a role. A couple of underlying trends explain this pattern, at least for the United States:

Increased interaction between friendly troops and friendly, neutral, or hostile civilian populations has become a feature of the contemporary operational landscape. This is likely to remain the case for the foreseeable future. Two factors account for this development. First, worldwide patterns of population growth and migration have resulted in increased urbanization, not only within the established industrialized states, but also in many undeveloped and developing societies. The urbanization of many crisis-prone regions of the world creates the potential for large, vulnerable groups of noncombatants to be caught up in military confrontations involving U.S. forces. Second, U.S. forces increasingly operate in the challenging environment known as military operations other than war. This category of operations includes such missions as humanitarian assistance, military support to civil authorities, peace operations, and noncombatant evacuations. These operations commonly involve close and continual interaction between friendly forces and noncombatant civilians.[27]

U.S. armed forces expect to find themselves "in a lot more combat in built-up areas, in cities, where collateral damage is a much more serious issue and avoiding civilian casualties is a much more serious issue than it has been in the past."[28] Close contact between soldier and civilian is often not smooth when the application of lethal force is involved.

The pervasive fear among both policymakers and the mass public of ever-spreading violence in this transformed global environment has—somewhat ironically—created a certain opposition to nonlethal weaponry. Even before the events of September 11, 2001, there was little doubt that greater public exposure to incidents of "terrorism, kidnapping, random acts of violence, urban unrest, increasing general crime, corporate crime, and weakened and poorly resourced and trained state law enforcement agencies" fueled these perceptions of insecurity.[29] Whether or not the actual level of violence, crime, or threat has in reality escalated is largely irrelevant, as improvements in domestic and international communication—continual reporting of incidents of unfathomable violence—has caused many people to live their lives in a state of perpetual terror and to demand permanent rather than temporary ways to stop the unruly sources of potential harm. This abysmal condition is at least as common and intense among people in advanced industrial societies as it is in the world's poorest countries.

Beyond the changes in the post–Cold War setting and the rampant feeling of danger, certain types of progress were necessary before technologically sophisticated nonlethal weaponry could become feasible.

More specifically, advances in precision-guidance technology have reduced the possibility that nonlethal weapons will hit unintended targets, and better command experience in peace support operations has fostered fixed operational requirements for the development of nonlethal weapons systems.[30] Indeed, in the context of potential use of nonlethal weaponry, "advances in information technologies, communications and precision-guided weapons are causing the military to reexamine its assumptions about how war can and should be fought."[31] Only in the last decade has this combination of human skill and technological capacity truly been realized. In some ways, what nonlethal weapons "really represent is the latest in a long line of weapons development based on the belief that advanced technology is the basis for military superiority."[32]

The United States has been the world leader in the development of nonlethal weaponry, an interest that became formalized in the late 1990s when the Pentagon began devoting "high-level" attention to nonlethal technologies.[33] Given that the U.S. Marines have consistently shown the greatest interest (along with U.S. Army MPs), it is not surprising that in 1997 the Department of Defense designated that the commandant of the Marine Corps run the newly created Joint Non-Lethal Weapons Directorate (JNLWD). With but a small, fifteen-person staff and a small, $25 million annual budget, this organization has made some progress; while some argue that it has not succeeded in providing sufficient overall coordination for nonlethal weapon development and policy,[34] the limited funds and staffing combined with significant external constraints have made its accomplishments seem remarkable so far. The topic of nonlethal weaponry has been the subject of considerable debate (of the type surrounding the introduction of any class of weaponry perceived to be new) within the U.S. government, both within the military establishment and more broadly among security policymakers from all branches of government.

The United States has already begun to deploy these weapons in foreign interventions and warfare. One example occurred during the 1991 Gulf War, when the U.S. Navy used a new class of highly secret nonnuclear electromagnetic pulse warheads carried on Tomahawk cruise missiles to disrupt and destroy Iraqi electronic systems, including their air defense systems. Another case occurred in early 1995, when U.S. forces returned to Somalia to safeguard the withdrawal of the remaining UN peacekeepers. The marines deployed there brought with them such weapons as guns that shoot rubber pellets and tiny beanbags to disperse crowds, stinger grenades that shoot rubber pellets, sticky foam that immobilizes people, and a foam barrier system laced with tear gas. General Anthony Zinni, the marine commander of the operation, expressed the belief that the adversary's knowledge that Americans possess these weapons helped to minimize casualties.[35] However, no systematically expressed doctrine or assessment of conditional utility has guided these applications. Decades earlier, of course, U.S. forces used substantial amounts of nonlethal defoliant herbicides during the Vietnam conflict.

More recently, the United States utilized nonlethal weaponry in the 2001–2002 war against terrorism begun in Afghanistan, though admittedly in a peripheral way:

> As the war on terrorism grinds on, U.S. military forces and civilian organizations are finding more and more uses for weapons that don't kill. . . . Marines guarding the newly reopened U.S. Embassy in Afghanistan, for example, are equipped with non-lethal rounds for their 12 gauge shotguns to drive away unarmed rioters. . . . U.S. troops overseeing al Qaeda and Taliban detainees at the naval base at Guantanamo Bay, Cuba, are training to use stingball grenades to put down a prison rebellion. . . . The Air Line Pilots Association International has called for the installation of stun guns as standard equipment in airline cockpits to thwart would-be hijackers with minimal risk to passengers.[36]

None of these applications pertained directly to combat action on the battlefield, but they have at least served a support function.

In the 2003 war against Iraq, in order to prevent or limit collateral damage, the United States needed to use "new nonlethal weapons that render power supplies, communications, and computers inoperable."[37] It is clear that "the massive media attention focused on the civilian casualties during the recent war on Iraq has again raised the question of how viable use of so-called non-lethal weapons may be."[38] According to Robin Copeland, an adviser on armed violence for the International Committee of the Red Cross, "In the history of warfare there has been a line in the sand drawn which is an attempt to keep out of the battlefield anything that involves toxicity on humans," but "the danger of the advances in technology that we're seeing now is that it might tempt us to step over that line."[39] Yet John Alexander, a former Green Beret and a leading advocate for the development of nonlethal weaponry, contends that "the changing nature of warfare, particularly since 11 September, required a change in tactics. . . . [T]he adversaries that we're facing now, particularly on the war on terror, are not ones who have signed any of these treaties [restricting the use of nonlethal weapons technologies]. [Y]ou are going to see these systems on the battlefield, and the battlefield is going to be defined quite differently."[40]

North Atlantic Treaty Organization (NATO) countries, in contrast, generally do not house nonlethal weaponry in separate directorates but rather fold it under their more general weapon systems acquisition mechanisms. These states have recently begun to grapple with the nonlethal weaponry question, signified in part by the first European symposium on nonlethal weapons, which took place in September 2001. International exchange agreements among the United States, Great Britain, and Israel, the NATO policy on nonlethal weapons adopted in September 1999, and recent U.S. and UK war games (dealing with urban warfare) have been exploratory first steps in revealing how and where nonlethal weapons can contribute to meeting war-fighting requirements.[41] Among major European powers, other unconventional weapons—particularly nuclear, chemical, and biological arms—have received far more sustained attention than nonlethal weaponry. Examples of European uses of nonlethal weaponry include Northern Ireland in the early 1970s, when the British army used nonlethal riot control weapons to contain turmoil,[42] and in the Falklands War of 1982, during which the British Royal Navy reportedly employed low-energy laser beams to distract Argentinean pilots attacking its ships.[43]

Nonlethal weapons development has spread recently to other parts of the world as well. For example, in 1995 China developed a portable laser disturber, designed to damage equipment and blind (at least temporarily) human targets; and the former Soviet Union has developed several laser weapons for air defense since the 1980s.[44] Nonlethal weapons are now even available through the international arms trade: "Compared to many high-technology conventional systems, non-lethal weapons can be marketed as relatively cheap alternatives and may even be sold at discounted prices in order to win new customers."[45]

Dangers of Nonlethal Weaponry

It is difficult to measure precisely the dangers of using nonlethal weaponry in modern warfare due to its limited employment in the post–Cold War security environment. Indeed, "the likelihood of non-lethal weapons making wars less brutal or destructive remains to be proven, [and] we don't yet know what the costs and effects of such weapons will be."[46] Regardless, there is a huge range of possible worries here:

> Keeping conflict minimally lethal requires a cooperative opponent and much good fortune. We must not only prevent him from killing, we need to prevent him from dying. Some opponents may find the goo and noisemakers a test of wills rather than our attempt to work around the problem of their survival and will begin shooting, killing, and dying, thus seizing the political initiative. Worse, the use of nonlethal alternatives may carry special penalties in terms of unrealistic expectations that a battle with non-lethal weapons will cause zero deaths. What merely stuns a 20-year-old soldier may easily kill a two-year-old child or a slightly out of shape 56-year-old professor. It also seems probable that the use of blinding laser weapons will be considered by the public as more horrible than the use of old-fashioned but quite deadly TNT. Here the weapon category "worse than lethal" may be appropriate. There are political as well as military minefields to cross in the quest to reduce war's killing.[47]

Chronicling specific possible threats seems useful to reinforce the imprudence of indiscriminate nonlethal weapons use.

One of the primary worries about nonlethal weapons, quite different from precision-guided munitions or information warfare, is that they potentially are inharmonious with international norms. Some proposed nonlethal weapons are not authorized under the international law governing weapons, and some nonlethal weapons may be deadly in some not-well-sanctioned employment scenarios.[48] More specifically, the International Committee of the Red Cross, Human Rights Watch, and other humanitarian organizations unambiguously "see such weapons, when used against civilians, as breaches of the Geneva Protocols."[49] Furthermore, many existing international treaties and weapons conventions do not specifically address this kind of armament.[50] The global legal system has unfortunately not kept pace with technological advances in weapons, and those weapon uses that circle around the edges of national and international laws can certainly pose problems for both legitimacy and effectiveness.

Some nonlethal technologies could be advantageous under certain conditions for enemies to use against advanced industrial societies. Several possibilities are conceivable. "For example, a terrorist group with rudimentary knowledge of our information switches could shut down our stock market with several well placed electromagnetic pulse generators."[51]

Western adoption of nonlethal technologies may create the risk that these nonlethal weapons will proliferate to hostile states and terrorist organizations.[52] Such a scenario might require the development of sophisticated countermeasures on the part of those very societies who invented nonlethal weaponry, societies who in some way are the most vulnerable to them because of the value placed on the well-being of soldiers and civilians. There is specifically a widespread fear that the United States is considerably more vulnerable to nonlethal weaponry than most of the countries against which it might be tempted to use them.[53] While such diffusion has not yet proven to be a problem, there is no guarantee that it will not emerge in the future, even though admittedly there is a clear mismatch between the advantages of nonlethal weaponry and the blatantly destructive philosophy and tactics of most unruly groups domestically and internationally. Finally, there is the possibility that foes targeted by nonlethal weaponry may inappropriately perceive it as a sign of weakness and become bolder in their unruly initiatives.

Concerns also emerge that unscrupulous governments could use nonlethal weaponry against their own people to suppress opposition. If such governments wanted to quell civil unrest, these coercive instruments could prove truly Orwellian in the scope of their insidious impact on domestic populations. Some skeptics feel that nonlethal weapons technology "can provide an authoritarian state with more means of oppressing and controlling people, and give police more tools for the abuse of power"; the possibility exists in these observers' eyes that the military under the employ of such a regime might use nonlethal weaponry "to bolster its own power and influence" and even to inflict torture on innocent citizens.[54] Of course, it is possible to misuse virtually any type of munitions in this way.

Outside of these applications by miscreants, the possibility always exists that inadvertently some innocent people will die or be severely and permanently incapacitated due to the use of nonlethal weaponry. The most vulnerable are usually the very young, very old, and sickly elements of a target population. A study released by the Pentagon's Defense Science Board in 1994 found that "a usually non-lethal weapon may cause unintended lethality under certain conditions: a stun gun could kill someone with a weak heart; a rubber bullet could hit a particularly vulnerable body part like the throat, and thus become lethal; and microwave devices could have unintended effects."[55] One type of nonlethal weapon, lasers designed to temporarily blind human targets, has generated special concern. In 1995, the Human Rights Watch Arms Project reported that the United States has pursued the development of at least ten different tactical laser weapons, many of which have the potential to blind individuals permanently.[56] Some nonlethal weapons can cause serious permanent damage or injuries that are worse—in terms of causing death after atrocious suffering—than those caused by lethal weapons:

> Super-glues and other adherents can suffocate people; EMP shockwaves can make helicopters crash; infra-sound can cause epileptic seizures in people; lasers can cause blindness; the new generation of acoustic weapons may merely be annoying, but they can also be intensified to cause shockwaves that could produce potentially lethal traumas. Plastic bullets are more readily lethal than rubber bullets—England had to recall 100,000 plastic bullets in 1999. Pepper spray has been a routine tool of the American police since 1987. As of 1998 it had caused at least 114 deaths, mainly due to conditional asphyxiation. Furthermore, non-lethal weapons can be used as instruments of torture. In California, police officers held the heads of ecological demonstrators motionless, opened their eyelids and put irritating liquid directly on their eyeballs.[57]

Critics also point to the 2002 siege of a Moscow theater by Chechen rebels, which Russian forces ended by using an opium-derived gas to overcome the hostage-takers. The gas, deemed nonlethal but never openly applied before, caused people to stop breathing and resulted in about 120 of the 800 civilian hostages dying from its effects. A Red Cross analyst said, "What we've learnt is that what is basically a medical agent that has been labeled non-lethal clearly carries a lethality."[58]

Beyond the physical damage, there is undoubtedly the danger of long-term psychological scarring among the victims of these arms. Skeptics remind us that being incapacitated and feeling intense pain from a source neither understood nor expected can be truly horrifying. These critics also point out that, while such an outcome may look humane compared to the killing associated with lethal weaponry, a different conclusion might emerge if the comparison point were instead presumably preferable noncoercive means of achieving security ends. Of course, even the staunchest advocates of nonlethal weaponry acknowledge this point as well as the real possibility of both death and deep psychological trauma.

More generally, nonlethal weaponry has the potential to diminish substantially the apparent horrors of war. More than other instruments of bloodless war, this form of armament can sanitize conflict in such a way that permanent suffering largely disappears. Widespread use of nonlethal weaponry could thus under limited circumstances make it more difficult to determine whether either side in a confrontation has won or lost, given that deaths on each side could be minimal, and thus prolong a conflict. And the use of these unconventional arms could also make relative power capabilities among adversaries harder to judge, leading to miscalculation about another's vulnerability.

The use of nonlethal weaponry can even conceivably foster unrestrained interventionism:

> A critical element of the debate is whether this represents a more effective means to manage crises or if it is a "slippery slope" to more frequent intervention in areas of marginal national interests or a mechanism promoting an escalation of conflict. The attractiveness of non-lethal weapons may drive decision makers to get involved because "we need to do something." The appeal of a low-risk, easy response may become addictive and thus cause inappropriate interventions and eventual military quagmires. There is no doubt that the availability of effective nonlethal weapons may provide an incentive for "adventurism."[59]

Indeed, several analysts contend that "the prospect of a relatively less bloody conflict may make the decision to intervene overseas more tempting to decision makers."[60] In many ways, the use of nonlethal weaponry can serve to legitimize otherwise unwarranted uses of force by minimizing the injuries involved.[61]

Finally, nonlethal weaponry could in some instances indirectly trigger a destabilizing new arena of international arms competition. In early 1995 mainland China became the first government in the world to openly market a blinding laser to other nations at an arms fair in the United Arab Emirates; this demonstrates the capability of other nations to develop and disseminate these weapons, and "thus, the prospect of a new arms race is a real possibility."[62] A race to develop expensive countermeasures could be stimulated as well: "As the non-lethal arsenal expands, threatened states will be driven to acquire protective or counter-measures to strategic non-lethal technologies."[63]

Conditional Utility of Nonlethal Weaponry

Despite these dangers, nonlethal weapons appear to have a very broad range of potential utility. For example, this type of armament could be useful in both urban and rural areas: "When fighting occurs in urban terrain, adversaries have a much greater opportunity to blend with the civilian population and/or hostages for protection from counterattack," increasing the desirability of weapons that do not kill; and "in a rural environment, non-lethal weapons may help reduce noncombatant casualties because of their inherent reversible effects on personnel and limited destructive effects on material property and the environment."[64] In general, some of the advantages possessed by nonlethal weaponry over lethal weaponry include creation or enhancement of a target's signature, countermobility and area denial effects, degradation of systems that produce and deliver weapons of mass destruction, deception, facilitation of movement over and through obstacles, capture of individuals for intelligence purposes, and protection of forces and facilities.[65]

Beyond immediate effectiveness during violent confrontations, many observers contend that nonlethal weaponry can aid in long-term deterrence, discouraging aggression when visibly postured and properly promoted. The Independent Task Force of the Council on Foreign Relations in July 1999 fully endorsed the findings of its 1995 predecessor that "non-lethal weapons possess the potential for providing new strength for diplomacy, give credibility for deterrence, and allow greater flexibility for the military, as well as create new strategic options for policymakers."[66] Others conclude that the application of such unconventional weaponry may end up "dissuading belligerents from resorting to more forceful weapons."[67] The basic reasoning is that deterrence is strengthened "by making potential adversaries aware that we can thwart aggression and achieve humanitarian aims in ways that do not entail prohibitive political costs, thus enabling us to act earlier, more freely, and more decisively."[68] The credibility of nonlethal weaponry derives at least in part from the assumption that there is a higher expectation that nonlethal weapons will actually be applied in conflict situations than lethal weapons due to the lower restraints—associated with the absence of excessive costs—on their use. Reducing the threshold for military actions could also in some cases allow conflicts to be contained earlier and with less bloodshed than if they were allowed to escalate due to exclusive possession of deadly force inappropriate for use in early stages of turmoil.

Clearly, however, this wide range of potential uses requires careful qualification. The propositions that follow are by no means exhaustive or comprehensive, and are certainly not devoid of controversy. Indeed, given the scarcity of recent wartime application of nonlethal weaponry, these conditions—which deal initially with characteristics of initiators and later with characteristics of targets—are even more tentative than those associated with the conditional utility of precision-guided munitions or information warfare.

First, and perhaps most obvious, nonlethal weaponry, for effectiveness, requires sound intelligence about intended targets. Indeed, to work properly, nonlethal weaponry may have significantly greater minimum intelligence requirements than traditional lethal weaponry.[69] One rationale for the importance of intelligence is that if the effects of nonlethal weapons are less visibly destructive, then military planners will face a special challenge, since they may not know if previous attempts have truly worked as they hoped. Intelligence on target damage seems particularly thorny:

> The nature of non-lethal "target damage" significantly complicates the assessment process. External indicators of success are not as obvious as the destruction caused by lethal munitions. The assessment of an embrittled bridge, acoustic incapacitating effects on personnel, or an EMP attack on air defense systems is not obvious from the traditional imagery-based assessments of military intelligence. The confidence in the successful attack on a target may not be confirmed until the enemy uses, or attempts to use, the particular equipment. In the case of air defense or offensive military equipment, waiting until friendly forces are engaged is too late to confirm disablement. Inaccurate assessments increase the risk to ground forces and air crews. Thus, non-lethal solutions may appear technically elegant but may not prove a credible capability unless the results can be confirmed.[70]

Developing new techniques of intelligence collection may thus be crucial for the proper functioning of this kind of weapon system.

Second, nonlethal weaponry may be more workable when applied by those with significant restraints on action based on humanitarian concerns about the sanctity of life. The justification for this claim revolves around this form of armament permitting coercive action to be taken without significant undercutting moral concerns about the loss of human life. Nonlethal weapons "can be an important tool in volatile situations where casualties, particularly noncombatant casualties, are unacceptable: for example, a squad handing out food at a distribution point would certainly prefer firing non-lethal crowd dispersal munitions instead of lethal weapons at an angry, violent crowd."[71] In a sense, nonlethal weaponry can respond to the increasing demand by democracies for "civilized conflict," minimizing casualties while still allowing for the pursuit of national interest in a manner that draws much less social criticism than traditional techniques of warfare.[72] Indeed, inside any democratic system, social acceptability appears vital: "Just as the basic decision to employ military force in defense of national interests is usually a matter of intense public concern, the manner in which that force is exercised is subject to the same scrutiny" because "some non-lethal weapons or their effects might—for religious or cultural reasons—prove so offensive to allies or important neutrals that their use would be counterproductive."[73]

Third, nonlethal weaponry appears to be more effective when used by those who possess superior overall military capabilities and combine these arms with traditional military and economic sanctions. Attempting to utilize this form of weapon without the presence of traditional coercive techniques makes nonlethal technologies seem hollow and, more important, lacking in communicated credibility about seriousness of purpose. It is widely held that "non-lethal application of force can be an effective deterrent or coercive tool in many circumstances" provided that the initiator "retains a credible threat of lethal force."[74] There is also widespread agreement that "non-lethal weapons must be complementary to current and planned conventional weapon systems,"[75] augmenting the use of deadly force rather than replacing it. This combination helps to ensure that a target's force calculations always account for superior destructive firepower on the part of the side initiating nonlethal weapons use. Moreover, adding economic sanctions to the mix can further reinforce the image of resolve, as nonlethal technologies could in some onlookers' minds "substantially increase the effectiveness of traditional sanctions and economic measures."[76] For example, the ability to force an enemy aircraft to withdraw from a prohibited airspace without shooting it down could be a valuable asset.

Fourth, nonlethal weaponry appears more likely to achieve its purpose when those employing it integrate it coherently into carefully formulated security objectives and broadcast clearly their security policy and possession of such arms. The underlying assumption here is that ambiguity about either possession or intended use hinders effectiveness in deflecting an adversary from aggressive plans. A combination of declaratory policy and visible possession of nonlethal capabilities is in accord with the more general principle that, for field effectiveness, "senior commanders must determine the rules of engagement and communicate them clearly and unambiguously."[77] Moreover, if the use of nonlethal weaponry appears ancillary rather than central in its relationship to security doctrine for management of the predicament at hand, or appears to be simply a function of possessing this weaponry rather than part of a carefully thought out strategic plan, then key advantages deriving from nonlethal weaponry could vanish. Because nonlethal weapons are often perceived as weaker than lethal weapons, and because less is generally known about these new technologies, the utility of these unorthodox coercive instruments rests more heavily on certainty about the will to employ them and operability in the field.

Fifth, nonlethal weaponry seems to work better when used against rational targets whose combatants fight for pay or duty, rather than against irrational targets whose combatants fight for passion or ideological cause. It seems reasonable to assume that nonlethal weaponry would have more of a dissuading impact on those who dispassionately calculate costs than on those who so devoutly pursue their cause that severe pain poses little hindrance to their plans (even though some nonlethal technologies in some situations make it physically impossible for any target—rational or irrational—to commit aggressive acts). Some observers contend that since "intense religious and political issues mark many of today's struggles," then "in these circumstances non-lethal weapons may not succeed as an effective coercive tool—non-rational factors may drive an irrational reaction or retaliation."[78] While nonlethal weaponry has the capacity to stop many adversaries in their tracks, what is in question here is the extrapolated transference of the notion of restraint—central to the concept of deterrence—from present to future predicaments.

Finally, nonlethal weaponry is more effective when used against targets engaging in small acts of spontaneous disruption, where targets pose acute short-term threats. This is not to say that nonlethal weaponry lacks broad strategic value, but rather to indicate that its greatest present utility is in tactical management of containable (in time and space) disruptive incidents. Indeed, a study conducted at the Center for Strategic and International Studies in late 1998, commissioned by the Office of the U.S. Secretary of Defense at the request of the National Security Council, admitted that "uncertainties in regard to the magnitude and duration of desired effects of nonlethal weapons, and in regard to the ability to minimize undesired effects, the potential for countermeasures, and the immaturity of the key technologies, makes it impossible, at present, to define a balanced program for strategic applications" of these coercive instruments.[79] Tactically, an opponent over time might be expected to develop a countermeasure to circumvent nonlethal weapons, even though the hope of those employing these technologies is that they will be robust in the face of potential countermeasures. For many long-run security problems, it appears that "non-lethal weapons would have little advantage over lethal systems, and the effect would only last as long as a physical presence was maintained."[80] Thus nonlethal weaponry constitutes a selective tool, rather than a dominant or primary tool, for use in coercive confrontations.

Conclusion

With the diverse and multifaceted forms of threat likely to emerge in coming decades, reexamination of the reliance on traditional modes of security is imperative. There is no question that an expanded array of coercive instruments could be potentially useful as a component of emerging casualty-minimizing techniques for managing conflict in areas where conventional military arms encounter severe limitations. Keeping in mind the largely tactical and defensive nature of nonlethal technologies thus far, "the relationship of non-lethal capabilities and the emerging strategic environment suggests that future non-lethal technologies could be decisive."[81]

This chapter has provided a more tempered understanding of when to include and when to exclude nonlethal weaponry as a means of casualty aversion in broader security strategies. The ultimate desired outcome would be the development of tighter guidelines and accountability for nonlethal weapons programs based on a continual reassessment of the costs and benefits of relying on nonlethal weaponry more heavily in the future. In the end, while the type of weapon employed has never been the sole important concern for determining the presence or absence of security, the notion of arms devoid of deadly force deserves our immediate and sustained attention.

Notes

1. This chapter draws heavily from Robert Mandel, "Nonlethal Weaponry and Deterrence Dilemmas," *Armed Forces & Society* 30 (forthcoming summer 2004).

2. Joint Non-Lethal Weapons Program, "1997: A Year in Review," unpublished document, Washington, DC, 1998, p. 2.

3. Joseph W. Cook III, David P. Fiely, and Maura T. McGowan, "Non-Lethal Weapons: Technologies, Legalities, and Potential Policies," unpublished

paper, November 10, 1997, available online at www.cdsar.af.mil/apj/mcgowan.html.

4. American Systems Corporation and Center for Naval Analysis, "Meeting the Demand of Future Military Operations: Study for Joint Non-Lethal Weapons Program," unpublished paper, Washington, DC, December 1999, p. 17.

5. Center for Defense Information, *Non-Lethal Weapons: War Without Death?* (Washington, DC: America's Defense Monitor, September 27, 1995), video.

6. Joint Non-Lethal Weapons Program, "Annual Report 1999," unpublished document, Washington, DC, 2000, p. 18.

7. Harvey M. Sapolsky and Jeremy Shapiro, "Casualties, Technology, and America's Future Wars," *Parameters* (summer 1996): 119–127.

8. Nick Lewer and Steven Schofield, *Non-Lethal Weapons: A Fatal Attraction? Military Strategies and Technologies for Twenty-First-Century Conflict* (London: Zed Books, 1997), p. 132.

9. Joint Non-Lethal Weapons Program, "1997: A Year in Review," pp. 1–2.

10. Center for Defense Information, *Non-Lethal Weapons*.

11. "New Military Technologies: Non-Lethal Disabling Weapons" (October 2002), available online at www.guerrasociale.org/nonletalweapons.htm.

12. Joint Non-Lethal Weapons Program, "1998: A Year of Progress," unpublished document, Washington, DC, 1999, p. 2.

13. Joint Non-Lethal Weapons Program, "A Joint Concept for Non-Lethal Weapons," unpublished document, Washington, DC, January 1998, p. 2.

14. John B. Alexander, *Future War: Non-Lethal Weapons in Twenty-First-Century Warfare* (New York: Thomas Dunne Books, 1999), p. 163.

15. Joint Non-Lethal Weapons Program, "1997: A Year in Review," p. 1.

16. Joint Non-Lethal Weapons Program, "A Joint Concept for Non-Lethal Weapons," p. 2.

17. Lewer and Schofield, *Non-Lethal Weapons*, p. 128.

18. Joint Non-Lethal Weapons Program, "A Joint Concept for Non-Lethal Weapons," pp. 2–3.

19. Richard L. Garwin and W. Montague Winfield, *Non-Lethal Technologies: Progress and Prospects* (New York: Council on Foreign Relations Press, 1999), p. 26.

20. Ibid., p. vii.

21. Ibid., p. 78.

22. Robert Mandel, *Armies Without States: The Privatization of Security* (Boulder: Lynne Rienner, 2002), chap. 1.

23. American Systems Corporation and Center for Naval Analysis, "Meeting the Demand of Future Military Operations," p. 1.

24. Cook, Fiely, and McGowan, "Non-Lethal Weapons."

25. Center for Defense Information, *Non-Lethal Weapons*.

26. Joint Non-Lethal Weapons Program, "1997: A Year in Review," p. 1.

27. Joint Non-Lethal Weapons Program, "A Joint Concept for Non-Lethal Weapons," pp. 1–2.

28. Center for Defense Information, *Non-Lethal Weapons*.

29. Alex Vines, "Mercenaries, Human Rights, and Legality," in Abdel-Fatau Musah and J. 'Kayode Fayemi, eds., *Mercenaries* (London: Pluto Press, 2000), p. 169.
30. Alexander, *Future War*, p. 6.
31. Center for Defense Information, *Non-Lethal Weapons*.
32. Lewer and Schofield, *Non-Lethal Weapons*, p. 131.
33. Center for Defense Information, *Non-Lethal Weapons*.
34. Garwin and Winfield, *Non-Lethal Technologies*, pp. 5–6.
35. Center for Defense Information, *Non-Lethal Weapons*.
36. Harold Kennedy, "U.S. Troops Find New Uses for Non-Lethal Weaponry," *National Defense Magazine* (March 2002), available online at www.nationaldefensemagazine.org/article.cfm?id=744.
37. Brad Knickerbocker, "War Aim: Quest to Reduce Accidental Casualties," *Christian Science Monitor* (March 14, 2003): 3.
38. Ibid.
39. BBC News, "The Weapons of Bloodless War," (May 13, 2003), available online at http://news.bbc.co.uk/1/hi/technology/3021873.stm.
40. Ibid.
41. Joint Non-Lethal Weapons Program, "Annual Report 1999," p. 1.
42. Lewer and Schofield, *Non-Lethal Weapons*, p. 59.
43. E. E. Cassegrande, *Non-Lethal Weapons: Implications for The RAAF* (Fairbairn, Australia: Air Power Studies Centre, 1995), p. 9.
44. Lewer and Schofield, *Non-Lethal Weapons*, p. 43.
45. Ibid., p. 44.
46. Center for Defense Information, *Non-Lethal Weapons*.
47. Sapolsky and Shapiro, "Casualties, Technology, and America's Future Wars," p. 122.
48. Cook, Fiely, and McGowan, "Non-Lethal Weapons."
49. Alex Salkever, "Time to Rewrite the Rules of War?" *Business Week Online* (April 1, 2003), available online at www.asia.businessweek.com/technology/content/apr2003/tc2003041_2114_tc047.htm.
50. Lewer and Schofield, *Non-Lethal Weapons*, p. 130.
51. Cook, Fiely, and McGowan, "Non-Lethal Weapons."
52. Joseph Siniscalchi, *Non-Lethal Technologies: Implications for Military Strategy*, Occasional Paper no. 3 (Maxwell Air Force Base, AL: Center for Strategy and Technology, March 1998).
53. Pat M. Holt, "Non-Lethal Warfare's Promises and Problems," *Christian Science Monitor* (August 3, 1995).
54. Lewer and Schofield, *Non-Lethal Weapons*, pp. 97–98, 133–134.
55. Center for Defense Information, *Non-Lethal Weapons*.
56. Ibid.
57. "New Military Technologies."
58. BBC News, "Weapons of Bloodless War."
59. Siniscalchi, *Non-Lethal Technologies*.
60. Center for Defense Information, *Non-Lethal Weapons*.
61. Brian Rappert, *Non-Lethal Weapons as Legitimizing Forces? Technology, Politics, and the Management of Conflict* (London: Frank Cass, 2003).

62. Center for Defense Information, *Non-Lethal Weapons*.
63. Siniscalchi, *Non-Lethal Technologies*.
64. Joint Non-Lethal Weapons Program, "1997: A Year in Review," p. 1.
65. American Systems Corporation and Center for Naval Analysis, "Meeting the Demand of Future Military Operations," p. 2.
66. Garwin and Winfield, *Non-Lethal Technologies*, p. 25.
67. Cook, Fiely, and McGowan, "Non-Lethal Weapons."
68. Garwin and Winfield, *Non-Lethal Technologies*, p. 26.
69. Edward P. O'Connell and John T. Dillaplain, "Non-Lethal Concepts: Implications for Air Force Intelligence," *Airpower Journal* 8 (winter 1994): 26–33.
70. Siniscalchi, *Non-Lethal Technologies*.
71. Joint Non-Lethal Weapons Program, "1998: A Year of Progress," p. 2.
72. Frederick Forsyth, "Send in the Mercenaries," *Wall Street Journal* (May 15, 2000).
73. Joint Non-Lethal Weapons Program, "A Joint Concept for Non-Lethal Weapons," pp. 9–10.
74. Siniscalchi, *Non-Lethal Technologies*.
75. Joint Non-Lethal Weapons Program, "1997: A Year in Review," p. 5.
76. Siniscalchi, *Non-Lethal Technologies*.
77. Alexander, *Future War*, p. 172.
78. Siniscalchi, *Non-Lethal Technologies*.
79. Garwin and Winfield, *Non-Lethal Technologies*, p. 37.
80. Alexander, *Future War*, p. 176.
81. Siniscalchi, *Non-Lethal Technologies*.

5
Information Warfare

Throughout the world, one of the biggest challenges is that of effective communication and management of complex information systems vital to national security.[1] It is easy to penetrate and jumble others' government information databases at the same time others are penetrating and disrupting yours, as well as to distort others' knowledge of what is transpiring through the introduction of manipulative and deceptive information. In a tightly interdependent globe, fragile but crucial data links develop among key states, based on vulnerable storehouses of electronic information, that are quite easy to disrupt in such a context. In recent decades there has been a growing obsession with the security of vital information banks and communications systems linked to national defense. Often called information warfare or "cyberterrorism," the nonviolent threats involved can undermine all the most critical authority structures.

Unruly rogue states and terrorist groups utilize such disruption because it increases the sense of anarchy conducive to their uncontrolled behavior and thwarts any effort to develop coordinated, coherent unilateral or multilateral strategies promoting security; and the great powers utilize such disruption and manipulation because doing so allows them to infiltrate and incapacitate potential threats from abroad, in some cases before these unruly groups can undertake any hostile action. Operating in secret and needing only a few resources, both status quo and non–status quo forces find that even the most complex information system is vulnerable to disruption, distortion, and eradication. Rationalizing that they are not causing any casualties but rather just demonstrating their cleverness and power by breaking computer security codes, this seemingly bloodless instrument can sometimes be prudent and yet at other times reflect both arrogance and irresponsibility.

Success on the battlefield is increasingly dependent on sound information management and securing the impermeability of information and communication systems. Modern warfare requires more than ever before speedy and reliable data on targets and coordination of multifaceted strategy and tactics in the field, and operating blind—even when possessing overwhelming force advantages—is a sure path to defeat. So even though quite disparate paths have emerged concerning how to achieve these goals, there is universal recognition—in a manner not present in the past—that information warfare is in many ways as critical as military preparedness, troop strength, and advanced weapons systems for success in modern warfare. Many analysts now argue that disrupting or distorting security information systems, rather than attacking traditional military targets, will be the primary thrust of future wars. If each system "becomes the center of gravity for modern militaries, it becomes the logical target of others," and for this reason "information warfare is often cited as the *leitmotif* of early 21st century conflict."[2] The incentives to disrupt one's enemies' defense information systems are identical to those for protecting one's own systems: "As everyone becomes increasingly dependent on automated information systems, the value of maintaining and securing them rises; conversely, the value to an adversary of gaining access to the system, denying service and corrupting its contents, also rises."[3] One analyst quips, "If you want to shut down the free world, the way you would do it is not to send missiles over the Atlantic Ocean—you shut down their information systems and the free world will come to a screeching halt."[4]

The security priority of information warfare has escalated rapidly. Central Intelligence Agency (CIA) director John Deutch announced in 1996 that cyberspace attack is one of the top threats to U.S. security, behind only the threat of weapons of mass destruction and the proliferation of nuclear, chemical, and biological weapons.[5] In 1996, President Bill Clinton called for an interconnected "cybersystem" that would provide early warning and minimize damage of attacks on computers controlling the stock market, banking, utilities, air traffic, and other "critical infrastructure" information banks.[6] In February 1999, CIA director George Tenet indicated that a key motive for focusing on this threat is that "terrorists and others are recognizing that information warfare offers them low cost, easily hidden tools to support their causes."[7] The George W. Bush administration has paid even more attention to information warfare, particularly after September 11, 2001, and has exhibited great "openness in regards to international broadcasting, military psychological operations and, most importantly, embedding international media on the battlefield."[8]

Compared to the other principal instruments of the quest for bloodless war—precision-guided munitions and nonlethal weaponry—information warfare is distinctive in a couple of ways. First, of all the means of minimizing the loss of human life during war, the techniques of information warfare have been subject to the most rapid change and most speedy global diffusion. Second, the way in which this approach contributes to casualty aversion is considerably subtler than the methods of the other instruments.

Information warfare is the only one of the three instruments that is widely available and heavily used by both the great powers and the unruly rogue states and terrorist groups. The impact of this situation is a potential stalemate: "As the United States moves toward using information warfare, so do its opponents; in fact, many say that the more the United States uses cyber-technology as a weapon, the more it exposes itself to cyber-attack by foreign governments, free-lance hacker/terrorists and clever cyber-criminals."[9] Information warfare has had considerable effectiveness when launched against the United States by its enemies. Although sophisticated computer network attack technologies may be very challenging for some, the low-tech information warfare toolkit is essentially open for anyone to use.

Furthermore, with information warfare the method of promoting casualty aversion is different. For precision-guided munitions and nonlethal weapons, the nature of the armaments themselves prevents widespread loss of life; while for information warfare, the relationship is more indirect, as to achieve victory this technique attempts to reduce the need for directly attacking and killing or injuring human beings. One interesting result from this more subtle connection between information warfare and minimizing loss of life is that, in cases where unintended deaths occur in the aftermath of a military confrontation, onlookers are less likely to be angry and resentful at what they perceive to be the failure of the technique than they would be in similar circumstances with precision-guided munitions or nonlethal weaponry.

This chapter first explores the wide scope of information warfare, including both information disruption and psychological operations; the motives for undertaking information warfare; the history of the wartime use of this technology; the dangers of relying on information warfare;

and finally its conditional utility in today's world. The aspiration here is to provide a more refined sense of when and how to employ information warfare as a coercive yet casualty-minimizing security instrument.

Definition of Information Warfare

Due to the recency of the heightened security concern about information warfare (sometimes more broadly referred to as information operations), an exceptional amount of ambiguity surrounds its definition, as there are many different facets of the offensive and defensive applications of this approach. The two major components discussed here, each of which associates tightly with the quest for bloodless war, are information disruption and psychological operations. Figure 5.1 summarizes the goals and methods associated with each component.

Information disruption is the most widely cited component of information warfare, and it includes corrupting, blocking, overwhelming, controlling, distorting, and leaking vital information or information systems in such a way as to endanger national or global security. A report prepared for the Defense Department titled *Information Warfare for Dummies: A Guide for the Perplexed* elaborates on this definition by including insertion of malicious code, theft of information, manipulation of information, and denial of service.[10] A major study of "cyberspace" threats more specifically details the spectrum of hostile information actions as including "inserting false data or harmful programs into information systems; stealing valuable data or programs from a system, or even taking over control of its operation; manipulating the performance of a system, by changing data or programs, introducing communications delays, etc.; and disrupting the performance of a system, by causing erratic behavior or destroying data or programs, or by denying access to the system."[11] The specific means by which these disruptions are undertaken include "computer attacks" whereby hackers gain access to a network as legitimate users and then proceed to steal good data, leave bad data behind, insert viruses, flood the system with messages (overwhelming it), or crash the system; all this can be accomplished with cheap equipment, readily available tools, and virtually nonexistent risk of detection.[12]

One important distinction exists between disrupting electronic communication and disrupting digital databases. The first involves stopping, distorting, or replacing messages sent across national boundaries or

Figure 5.1 Information Disruption Versus Psychological Operations

Information Disruption	Psychological Operations
Goals	
	Prevent effective functioning of command-and-control systems within states. Reduce human casualties.
Corrupt, incapacitate, or distort vital information systems.	Convey selected information to foreign audiences to influence their emotions, and motives to undermine their confidence in leaders, and, ultimately, change the behavior of foreign governments.
Disrupt communication among enemies as well as their information systems.	
	Control the flow of information and how it is interpreted, ultimately controlling how people think.
Methods	
Insert malicious code or false data, steal or destroy crucial data or programs, manipulate information, gain control of information systems, cause erratic data behavior, introduce communication delays, engage in electronic decapitation and sensor denial, or deny system access.	Spread propaganda and deceptive images using cognitive stereotypes and simplifications, often exploiting ethnic, social, or moral cleavages in the target society.
	Drop leaflets, send radio messages, and leave e-mail and cell phone messages supporting one's actions and discrediting those of one's enemies.
Use computer attacks where hackers gain access to a network and then proceed to steal good data, leave bad data behind, insert viruses, flood the system with messages, or crash the system.	

between military commanders and their troops in the field, while the second involves corrupting or modifying the basic information used by government defense agencies to conduct their security operations. More specifically, the span of information reach is quite vast: it can address data themselves or the vital communications systems in which data are embedded; it can remove, distort, or add data; and it can eliminate or distort communication, add noise to a communication system, or even expand communication to include undesired parties.

Another important practical differentiation exists between the largely offensive use of information disruption, undertaken mainly by rogue states and terrorist groups, and the largely defensive use of information disruption, undertaken mainly by advanced industrial societies.

It is clear that in today's tightly networked world, information systems may serve equally well as both weapons and targets.[13] These uses are not mutually exclusive, as the United States engages in both offensive and defensive strategies. On the one hand, offensive "information warfare is a veritable option for the U.S. to employ to advance its foreign policy interests; as the pre-eminent information society, the United States possesses the technological knowledge to wage an effective information war."[14] On the other hand, with regard to defensive information warfare, the defense and intelligence community has "done a commendable job in identifying and adjusting to the new national security threat posed by information warfare" initiated from abroad.[15] Thus, "far from merely a helpless victim, the United States has also exploited the insecurities of the networked world."[16]

Aside from information disruption, the second vital component of information warfare is psychological operations, which involve spreading propaganda and deceptive images. The Defense Department defines psychological operations as "planned operations to convey selected information and indicators to foreign audiences to influence their emotions, motives, objective reasoning and, ultimately, the behavior of foreign government, organizations, groups or individuals."[17] In attempting to manipulate and exert mind control over the target, "one side may try to demonize the other by using cognitive stereotypes and simplifications while making their own side appear just."[18] Ironically, democracies' emphasis on freedom of speech often causes psychological operations aimed at foreign targets—whether overt or covert—to garner more legitimacy than most forms of information disruption.

Despite the changing nature of information warfare, linked to the evolving state of information technology, it is possible to obtain at least a general sense of its parameters. Information warfare includes internal as well as external interference, and such disruption can be unintentional as well as intentional. Information disruption focused on incapacitating a state regime's electronic databases, or psychological operations focused on distorting a mass public's perceptions, can work hand-in-hand to interfere with and undermine target government authority. As immediate accurate electronic communication becomes increasingly essential to the multifaceted and interdependent command-and-control structures of modern military defense systems, even a modest form of information disruption or propaganda that simply involves a slight delay, modification, or contamination in the transmission of vital communication or data can be sufficient to paralyze an entire security structure and make it unable to manage or cope successfully with an array of complex threats.

Motives for Information Warfare

Information warfare serves to prevent the effective functioning of command-and-control of military operations within states while at the same time preventing accurate military-coordination communication from flowing across states. Its relative invisibility to outside observers makes it all the more attractive for use. The U.S. defense community is interested in information warfare for two reasons: its own vulnerability, as "the United States, in civilian as well as military matters, is more dependent on electronic information systems than is anyone else in the world"; and its enemies' vulnerability, as information warfare "may be as much an opportunity as it is a threat."[19]

The perpetrators of information warfare are quite diverse, as these activities require no more than "a powerful computer, a keen mind, and an underlying grudge."[20] The deviant assortment includes "hackers, zealots or disgruntled insiders, to satisfy personal agendas; criminals, for personal financial gain, etc.; terrorists or other malevolent groups, to advance their cause; commercial organizations, for industrial espionage or to disrupt competitors; nations, for espionage or economic advantage or as a tool of warfare."[21] When the goal is social status—such as when initiators are just young computer hackers who desire to test their virility by breaking into the most secure and protected data systems—the danger is in many ways the lowest. When the objective is economic, the security threat is more ominous. For example, some want to promote anarchy by using information disruption to break down the ability of international financial networks or government military institutions to function, while others—particularly transnational criminal organizations—are motivated by greed, and use information warfare to gain inside information or alter databases for financial benefits. When the objective is political or military, the challenge posed to the state seems the highest. For example, some use this technique to attempt to alter the policies of another government or its level of support from its people, while others seek to prevent soldiers from receiving proper command directives during warfare. Often the goals combine, with perpetrators disrupting target command-and-control capabilities to organize, assess, and disseminate data and at the same time fostering fear and dissention among target soldiers and civilian populations. Attempts to protect against these threats or negate others' capacity to engage in information warfare are just as common, usually undertaken by state governments or major corporations rather than by individuals or small groups.

The low cost of information warfare and the ubiquitous availability of its techniques also add to its appeal as a security instrument: "Offensive information warfare is attractive to many because it is cheap in relation to the cost of developing, maintaining, and using advanced military capabilities; it may cost little to suborn an insider, create false information, manipulate information or launch malicious, logic-based weapons against an information system connected to the globally shared information infrastructure."[22] The generally inexpensive nature of these techniques stands in stark contrast to the cost of the other two primary instruments of casualty aversion—precision-guided munitions and nonlethal weaponry—and thus partially explains the relative omnipresence of information warfare among all sorts of states in today's world.

More narrowly, two contrasting critical motives behind information disruption are electronic decapitation and sensor denial. Electronic decapitation works in a similar fashion to leadership removal:

> Deny enemy command and control elements the use of any automated or electronic decision aids. This constitutes "electronic decapitation." Data bases, data fusion systems, electronic processing and display systems for command centers, combat information centers, and the like must "go dark." Introduce "combat amnesia" to the enemy. . . . Cut or deny all the enemy's information-transfer media—telephone, radio frequencies (RF), cable, and other means of transmission. Sever the nervous system. Deny, disrupt, degrade, or destroy every transmission.[23]

Sensory denial is parallel to electronically eliminating a foe's five senses:

> Kill sensors, not people, first. Open the way to the enemy's army by blinding all his defenses. Deny electronic radiation. If it radiates, it dies. Seek absolute silence over the battlefield. Homing weapons, jamming, and lethal and nonlethal suppression of enemy defenses must be employed. . . . Overpower passive sensors. Burn passive detectors. Use lasers or optical trackers. RF receivers should be blown.[24]

These means are extremely effective against a variety of high-tech foes.

Looking more exclusively at psychological operations, the underlying principle is to control the flow of information and how it is interpreted, ultimately determining how people think:

The target of netwar is the human mind. One could argue that certain aspects of the cold war had the characteristics of a dress rehearsal for future netwar. Consider, for example, Radio Free Europe, the Cominform, Agence France Presse, or the US Information Agency. But netwar may involve more than traditional state-to-state conflict. The emerging of nonstate political actors such as Greenpeace and Amnesty International, as well as survivalist militias or Islamic revivalists, all with easy access to worldwide computer networks for the exchange of information or the coordination of political pressure on a national or global basis, suggests that the governments may not be the only parties waging Information War.[25]

Indeed, on any given international security issue, governments often are neither alone nor the most adept at using propaganda against targets to get them to alter their beliefs. More specifically, psychological warfare is "waged against a general population in order to undermine confidence in leaders or the wisdom of their actions, often exploiting ethnic, social, or moral cleavages in the target society."[26] Its purposes include to "amplify effects of military operations, inform audiences in denied areas; overcome censorship, illiteracy or interrupted communication systems; give guidance or reassurance to isolated, disorganized audiences; target opponent audiences to diminish morale or will to resist; sustain morale of resistance fighters; exploit ethnic, cultural, religious or economic differences; give opponents alternatives to conflict; influence local support for insurgents; support deception operations; project favorable US image; and use all available means to channel the target audience's behavior."[27] Recent technological developments allow psychological operations to be customized for particular target audiences. Like information warfare more generally, psychological operations are "designed to achieve almost costless advantages over one's adversaries";[28] but unlike most other forms of information warfare, psychological operations can often be carried out in a very low-tech manner without the use of sophisticated electronics.

Generally, most analysts agree that information disruption reduces the occurrence of casualties because it targets the communication and database infrastructure rather than human enemies: strategic information warfare "seems to possess the redeeming quality of being 'much more humane' than other forms of strategic warfare since the only intended casualties would be the crippling of information flow, convenience, and comfort."[29] Moreover, the increased dependence on communication and computer systems, combined with long-range precision delivery systems and new "nonlethal" methods of electronic warfare, has in several instances "greatly limited" casualties.[30] Of course, information disruption efforts may easily fail or even backfire.

Similarly, the express purpose of psychological operations during warfare is "to prevent needless loss of life, needless casualties."[31] This approach may soften the repercussions of war, as such operations "can reduce casualties by encouraging opposing troops to surrender, and they can help win civilian support."[32] Indeed, victory may rest more on control of foes' attitudes than proficiency in combat against them:

> We can discern a goal for information-age psychological operations (PSYOP)—"to compel the enemy to do our will without fighting." This goal is particularly relevant today in view of an increasing American intolerance for casualties. Information-age PSYOP, more than any other military instrument, may provide us with an increased capability to pursue our national interests without bloodshed.[33]

Reminiscent of the concept of mind control, if effective this technique could destroy the morale of enemy troops. However, the power of propaganda faces some clear limitations in today's world:

> Propaganda, unfortunately, has frequently been of only limited utility. It has been used since the dawn of organized warfare in both a positive and negative sense. It has always been designed to either inspire confidence in one's own people and leaders and to alternatively ridicule, frighten, or demonize one's enemy. As such, it has always occupied a supplemental place in war, but that is all. . . . The ultimate problem with even the slickest propaganda is that it does not always work, and even when it does, its effectiveness is limited.[34]

Moreover, as with information disruption, psychological operations can certainly backfire:

> For example, during the first Gulf War, Iraqi officials tried to disillusion U.S. soldiers with a broadcast of a woman, Baghdad Betty, warning them that their wives and girlfriends were being seduced by actors like Tom Cruise, Tom Selleck and Bart Simpson while they were away. But, for obvious reasons, American soldiers weren't adversely affected by the misguided—ultimately rather comical—operation.[35]

Furthermore, if a state discovers and publicly discloses attempted deception or distortion by a foreign power on a critical matter, the political consequences could be dire.

This wide range of motives and purposes surrounding information warfare suggests the difficulties of detecting unambiguous success from its usage. As one military analyst quips, "When you're launching a computer attack against somebody, how do you know you've got them and haven't hurt yourself?"[36] A common notion is that "victory in information warfare depends on knowing something that your adversaries do not and using this advantage to confound, coerce, or kill them; lose the secrecy, and you lose your advantage."[37] Yet sometimes you can accomplish one information warfare objective only to inadvertently thwart another. For example, "should the perfect information warfare campaign leave the enemy command with no means to fight, it is then nonetheless unable to communicate its desire for surrender or truce to its troops."[38] In a highly interpenetrated world, information warfare copes with both crosscutting attempts to disrupt information systems and public and private propaganda organs pumping out falsehoods; so if there is change within a target following an information warfare initiative, it is difficult to know whether it was the initiative that resulted in this change.

Information Warfare in Recent History

Two "revolutions" are involved in the historical origins of incapacitating information disruptions: the revolution in military affairs, which has turned attention away from traditional declared wars with standing armies to new forms of unconventional conflict; and the information revolution, which has dramatically increased our dependence on data and computers to prepare for and fight wars. The result of these two developments is that "warfare is no longer primarily a function of who puts the most capital, labor and technology on the battlefield, but of who has the best information about the battlefield."[39] Put another way, defense analysts often believe now that "warfare is less and less about pushing men and machines around the battlefield and more and more about pushing electrons and photons."[40] While "information has been associated with power, war, and the state since at least the time of the Greek gods,"[41] the global openness and free access to information in the post–Cold War environment is unprecedented in human history, and as a result seems likely to cause "substantial discomfort" through its highlighting of the tensions between security and freedom.[42] Moreover, with the flood of information, "power is migrating to small, nonstate actors who can organize into sprawling networks more readily than can traditional hierarchical nation-state actors."[43] Sophisticated information and communication systems have improved the way major powers fight wars but at the same time have made military command-and-control more vulnerable to attacks.[44]

For many, the impact of the information revolution in particular on national security has been monumental:

> The revolutionary agent here is the microchip, which, combined with strategic and organizational innovations, is said to be transforming warfare in the same way the musket did in the 1600s and the atom bomb did in 1945. Advocates foresee a "digital battlefield" that weds precision-guided long-range weapons to high-tech information and surveillance systems that enable commanders to direct the action from thousands of miles away, and so on down to the smallest computerized gizmo that springs from the mind of a military planner.[45]

Government officials are now talking about utilizing "weapons of mass communication" instead of "weapons of mass destruction." How much of a discontinuity this change creates is still a matter of debate.

Information warfare of all sorts has been used throughout the ages, with its application escalating particularly after electronic systems were introduced to security, and recently it has become more refined and often explicitly associated with casualty aversion. For the United States in the 1990s, information warfare developments moved beyond simple defensive electronic countermeasures into more sophisticated offensive efforts to influence and manipulate enemy information systems.[46] Despite the secret nature of most incidents, a large number of anecdotally related cases of information disruption serve to illustrate the growing historical importance of information warfare. During the U.S. Civil War, "when signal flags were used, Union forces broke Confederate coding schemes and diverted the South's troops by planting bogus messages."[47] Perhaps the most famous twentieth-century case occurred during World War II: Winston Churchill employed information warfare when he utilized the Enigma machine (the machine used by the Germans to encipher high-grade wireless traffic) to read German codes. This famous success inspired many later efforts like it.

Similarly, for centuries national governments have engaged in perception manipulation of their own citizenry and other countries' leaders, but recently the techniques of psychological operations have become much more sophisticated. A comparison of how the United States used propaganda to gain support during the 2003 war against Iraq and how the Germans under Goebbels used propaganda to gain support for the Nazis during World War II shows great advances in sensitivity to the target audience and their desires and fears. Indeed, it is worth noting that concern about the successes of German propaganda first prompted the U.S. government under the Roosevelt administration to look into international broadcasting in 1940, resulting in the creation of Voice of America.[48] In 1947 Voice of America switched its target to the Soviet Union, followed by Radio-Free Europe in 1950 and Radio Liberty in 1951. Of course, the Japanese used the seductive "Tokyo Rose" during World War II to try to undermine U.S. soldiers' will to fight.

Specific post–Cold War information disruption incidents with the United States as a target include Dutch hackers penetrating the Pentagon's computers and offering to sell troop-movement plans to Iraq during Operation Desert Storm in 1991; Russian criminals illegally transferring huge sums of money from Citibank into accounts abroad in 1995; German hackers selling classified U.S. intelligence to the successors of the KGB in 1995 and 1996; and hackers who shut down the Central Intelligence Agency website in September 1996 to make a political statement.[49] In February 1998, two Californian teenagers (aided by an Israeli teenager) gained access to numerous Pentagon information networks, a disruption Deputy Secretary of Defense John Hamre termed "the most organized and systematic attack" on U.S. defense networks yet discovered.[50] Of course, the United States is by no means alone as a target: two illustrative, widely reported incidents are the revelations in June 1996 that several London financial institutions had been paying for three years huge sums to international criminals who had penetrated their computers and convincingly threatened to shut down the systems unless they received the money; and the report in June 1998 that a group of hackers from the United States, England, Holland, and New Zealand had managed to break into India's national security computer network and steal sensitive nuclear weapons secrets. These disruptive incidents clearly represent only the tip of the iceberg.

More generally, the random and sporadic nature of these information break-ins, augmented by the highly classified nature of their overall pattern, makes it impossible to estimate accurately the aggregate global frequency and impact of this ominous transnational activity. However, the experience of the U.S. military, which has more than 2 million computers and more than 10,000 local area networks, gives a troubling indication of the pervasiveness of the problem. In 1995 alone the Pentagon logged more than 250,000 attacks on nonclassified computer systems, and in more than 60 percent of the cases the hackers succeeded in penetrating the system.[51] The costs of this kind of disruption are measured not only by the damage done to the information systems but also by the money spent on countermeasures; one sign of the high toll here is the 1997 recommendation by the Defense Science Board that the Defense Department augment its current information warfare budget of less than $1 billion with an additional $3 billion over the succeeding five years, largely for defensive purposes.[52] More recently, the U.S. Government Accounting Office estimated that 120 groups or countries have or are developing information warfare systems; and the Center for Strategic and International Studies claims that twenty-three nations have cybertargeted the United States.[53]

Utilizing information warfare to minimize casualties was a key component of the 1991 Gulf War:

> In the Gulf War a major contributing factor to keeping friendly casualties low was the coalition's ability to blind the Iraqis' operational surveillance and reconnaissance capacity. A blind opponent doesn't know where to counter-strike. The campaign to blind one's opponent while seeking to optimize one's own surveillance and reconnaissance has been termed "information war." It is characterized by attempts to extend and enhance one's information acquisition, processing, and communication capabilities while degrading or destroying those of the enemy. It also requires mounting countermeasures to block the opponent's efforts to disrupt and degrade one's own capability. The U.S. and its allies have aggressively pursued superior information war capabilities for the last 25 years in a historical context of competition with the Soviets. The rationale has been that the West could overcome Soviet quantitative advantage by putting Western forces and firepower in the right place at the right time while simultaneously causing the disruption and disorganization of the operations of the larger Soviet forces.[54]

Psychological operations "helped persuade thousands of Iraqis through leaflets and loudspeakers to surrender during the Gulf War; in one case, 500 Iraqis left their bunker to give themselves up to three psyops soldiers armed with bullhorns."[55] Moreover, "inducing large numbers of Iraqi soldiers to defect or surrender" through radio broadcasts over Iraqi-held Kuwait "helped prepare the battlefield" for U.S. success.[56] In November 1990 the United States began broadcasting Voice of America into Kuwait, and "these radio broadcasts played a large part of the psychological operations (psyops) that helped to prepare the battlefield by offering the Iraqi soldiers food, bedding and medical care if they surrendered and reminded them of the consequences if they did not"; and in January 1991 the United States adopted the Voice of the Gulf, which began nonstop radio broadcasts from both land-based transmitters in Saudi Arabia as well as nearby aircraft.[57] Information warfare efforts during this conflict appeared to be especially significant because "although the use and exchange of information have been critical elements of war since its inception, the Gulf War was the stage for the most comprehensive use of information, and information denial, to date."[58]

In the 1999 Kosovo conflict, the United States again initiated concerted cyberattacks. Despite the Defense Department finding that information warfare initiatives in Operation Allied Force largely failed, U.S. Air Force general John Jumper and U.S. Army general Hugh Shelton "confirmed that a successful information warfare (IW) attack was carried out to confuse and disable the Yugoslav air defense system";[59] more specifically, "false messages and targets were injected into Yugoslavia's complex computer-integrated air defense system."[60] With the help of neighboring nations, the United States used Radio Free Europe and Radio Liberty to create a "ring around Serbia" in which "U.S. international broadcasting greatly contributed to the fall of Milosevic in October 2000."[61]

Similarly, information warfare played an important role in the war against terrorism in Afghanistan. The strategy was "to disable air defence systems, scramble enemy logistics and perhaps infect software," with the underlying goal being to "wreak electronic havoc on countries accused of harbouring terrorists."[62] In addition, specialists in psychological operations, "armed with mobile broadcast stations, leaflets and loudspeakers," sought to "demoralize and strike fear in the Taliban while bucking up refugees and convincing Afghans that Osama bin Laden, not the United States, is their enemy."[63] Within weeks of the September 11, 2001, terrorist attacks in New York and Washington, D.C., U.S. radio broadcasts began over the airwaves in Afghanistan, including an explanation "that the attacks in New York and the Pentagon were on innocent people—an act forbidden by the Muslim Koran."[64]

Finally and most recently, the 2003 war against Iraq exhibited more of the same kinds of information warfare applications:

> The US is stepping up psychological operations, or psy-ops, to destabilise the Iraqi regime in the run-up to the start of military action. These

include the Information Radio broadcasts carried since December 2002 by US Lockheed Martin EC-130E Commando Solo aircraft flying over southern Iraq. The USA is also dropping millions of leaflets over Iraq telling the population where to find these broadcasts on their radio dial.[65]

This may be the most concerted information warfare effort in history:

> In the most ambitious effort of its kind, the American military is already at war with Iraq, but it is a conflict being fought with electrons and words in advance of any order by President George W. Bush to loose bullets and bombs. American cyber-warfare experts recently made an e-mail assault against Iraq's political, military and economic leadership, urging them to break with the regime. A second wave of messages has gone to private cell phone numbers of specially selected officials. More than eight million leaflets have been dropped over Iraq, including towns 100 kilometers south of Baghdad, warning Iraqi anti-aircraft missile operators that their bunkers will be destroyed if they track or fire at allied warplanes. A similar blunt notice has gone to Iraqi ground troops: Surrender, and live. Radio transmitters hauled aloft by Air Force Special Operations EC-130E planes are broadcasting directly to the Iraqi public in Arabic with programs that, by mimicking the styles of local radio stations, are generations advanced from the clumsy preachings of previous wartime propaganda efforts.[66]

Beyond these techniques, the U.S. information warfare operation in Iraq included the sending of e-mails to prominent Iraqi officials to offer them clemency after the fall of the regime, in return for assistance in finding weapons of mass destruction; and swamping mobile phones of senior officials close to Saddam Hussein with calls urging them to disobey orders. The radio broadcast operations specifically included Radio Tikrit, which initially pretended to be a pro-Saddam station but after two weeks radically changed its tone and sharply criticized the Saddam regime and urged Iraqi soldiers to defect; Radiyo al-Ma'ulumat (Information Radio), a U.S. propaganda operation that also broadcast anti-Saddam messages; and Radio Free Iraq, an Arabic-language service indirectly beamed into Iraq from Radio Free Europe/Radio Liberty headquarters.[67]

By all estimates, this application of information warfare during the 2003 conflict in Iraq had a substantial impact. The purpose of this effort was to reduce the carnage during war, bring about a swifter and less painful end, and improve the political aftermath of the conflict:[68]

Military planners at the United States Central Command are using the burgeoning field of information warfare—including electronic attacks on power grids, communications systems and computer networks, as well as deception and psychological operations—to try to break the Iraqi military's will to fight and sway Iraqi public opinion. Commanders may use supersecret weapons that could flash millions of watts of electricity to cripple Iraqi computers and equipment and literally turn off the lights in Baghdad. "The goal of information warfare is to win without ever firing a shot," said James Wilkinson, spokesman for the Central Command in Tampa, Florida. "If action does begin, information warfare is used to make the conflict as short as possible." Deception and psychological operations have been a part of warfare for centuries, and American commanders launched limited information attacks during the 1991 Gulf War and 1999 air campaign over Kosovo. Commanders say the current effort is much broader and more tightly integrated into the main war plan than ever before. "What we're seeing now is the weaving of electronic warfare, psyops and other information warfare through every facet of the plan from our peacetime preparations through execution," said Major General Paul Lebras, head of the Joint Information Operations Center.[69]

Military deception and information overload were also part of the package designed to keep the adversary off balance.[70] As a key component of the psychological campaign, "a massive propaganda blitz preceded the start of ground operations," in which "U.S. forces used leaflets, radio broadcasts, faxes, e-mails, and other means to urge Iraqi troops not to fight"; while this campaign did not quell all resistance, "it contributed to the decision of most regular [Iraqi] army units to stay out of the fray."[71] Overall, "experts generally give high marks to the way the U.S. military has showered Iraqi soldiers with reasons to surrender and tried to sow internal dissension among Iraqi leaders."[72] Nonetheless, the success of Al-Jazeera and other pro-Iraqi Arabic satellite news channels showed the limits of the U.S. capacity to dominate the airwaves completely.[73]

Dangers of Information Warfare

On an individual level, there are several different types of fears associated with the use of information warfare. First, knowledge, manipulation, and falsification of personal data—whether it be related to identification, finances, or criminal record—is a real concern. The "concept of information warfare is very dangerous from a civil liberties point of view," entailing that "in order to ensure our survivability in an information war, the military should make use of all 'national assets and use all sectors of society,'" including "all privately owned computers, fax machines, computer bulletin boards and . . . even the assets of international corporations."[74] Second, the elimination of any truly credible sources of information, possibly due to extensive psychological operations, can create panic and confusion among the public, eliminating the informed voice of the citizenry that is so necessary for the proper functioning of democracy. Third, the sense of personal and national vulnerability can be heightened, as the recognition has grown that—even with sizable military forces—no way really exists to protect oneself and one's society from penetration by electronic disruptions and distorting propaganda.

Along with these fears, information warfare poses a simultaneous danger of desensitizing those who use it to the serious human costs involved:

> While an information attack may avoid direct human casualties there may be considerable indirect death and damages. Disrupting the information infrastructure of another nation will shut down hospitals, cause planes and trains to crash, cause starvation in isolated regions, etc. Though there are no direct casualties when logic bombs destroy the information infrastructure of another nation, they may cause significant collateral death, most likely civilian.[75]

Taking action in cyberspace may mask the true extent to which harm to people is occurring as a result. For example, the disruption of a state's power infrastructure—its ability to provide oil, electricity, and gas for consumer and corporate energy needs, industrial production, and transportation—may cause some innocents affected to freeze to death; or a cyberattack on a computer server may prevent a hospital from undertaking complex surgeries on near-terminal citizens. Many forms of information disruption cannot readily discriminate in their negative effects between combatants and noncombatants. Thus, indirectly, information warfare may occasionally cause as much human damage as traditional warfare.

For both individuals and states, an uncontrollable flow of massive quantities of sensitive information across national borders can be just as debilitating as having that flow interrupted. While most of media and government attention has been directed at information disruptions caused by efforts to hack into important systems and disabling them, the specter of destroying a classified defense information system by leaking out critical pieces of data would be just as damaging, has probably occurred with about equal frequency, and is just as difficult to safeguard. Either way, national security is severely compromised.

Perhaps the greatest danger posed for states that make extensive use of information warfare as part of a casualty minimization strategy in military confrontations overseas is that they themselves will increasingly become targets of the same kind of penetration and propaganda. In the words of one analyst, "There is also the genie-in-the-bottle syndrome to think about," as "once a cyber-attack has been unleashed, who's to say that in the interconnected world your carefully constructed virus won't spread to the networks of friendly or neutral nations?"[76] An alarmist 1995 *Washington Post* article warned of the threat of an "electronic Pearl Harbor" emanating from the spread of information disruption capabilities, in which, "if the civilian computers stopped working, America's armed forces couldn't eat, talk, move or shoot," and would have "no ability to protect themselves from cyberattacks and no legal or political authority to protect commercial phone lines, the electric power grid and vast databases against hackers, saboteurs and terrorists."[77] Obviously if great powers utilize this technique frequently, which they do, this inadvertently legitimizes its use by others as a means of achieving their often contrary foreign security policy ends. Among other problems, everyone could end up experiencing data overload: "The danger now is that commanders will be so bombarded by a blizzard of largely extraneous or even unessential data that it will obscure the real issues that have to be dealt with," making it nigh impossible to sort out important truths through extraneous noise.[78] This could lead to escalating information warfare battles that would serve nobody's interests and throw societies into complete chaos.

Possible increased diffusion of sophisticated information warfare technologies to one's foes reinforces this specter of information warfare retaliation. Many of the techniques are so inexpensive and easy to implement—far more so than the other instruments of casualty aversion—that anyone could readily copy and apply them, from hacking into government websites or databases to launching disinformation campaigns: "The combination of low cost, widespread availability, and lack of controls makes the tools for waging digital warfare highly accessible."[79] Not only rogue states but also nefarious transnational groups could begin to use more of these techniques to disrupt global stability. For example, outside of the obvious case of terrorist use, when transnational criminal organizations undertake such activity, they "pose grave threats to the integrity of the world financial system, undermine the ability of states to protect their citizens, and are themselves a major threat to human rights": this "illicit financial community" steals about $10 billion a year from U.S. financial institutions alone, and with ever-expanding monetary resources "will capitalize on its current strategic advantage and provide a greater challenge to the stability of financial institutions worldwide" by using the Internet to move money quickly to virtual banks in offshore havens outside the reach of national authorities.[80] These criminal activities represent a classic case of distortion, manipulation, and even outright falsification of data that is highly disruptive to the functioning of civil society in a more bottom-up way than intrusions into defense communication systems.

The widespread use of information warfare could accelerate the sophistication of emerging techniques so rapidly that offense might very well outstrip defense, meaning that no state would be able to protect itself from intrusion from the outside. The vulnerability of Western advanced industrial societies is particularly noteworthy: their financial databases could be penetrated, communications jammed, and their energy and transportation infrastructure garbled through mechanical, electromagnetic, and digital information warfare attacks.[81] This disruption is extremely difficult to deter because attackers may be anonymous and, even if perpetrators are identified, it is difficult to develop effective retaliatory options.[82] Because "defense, the police, banking, trade, transportation, scientific work, and a large percentage of the government's and the private sector's transactions are on-line," the result is exposure of "enormous vital areas of national life to mischief or sabotage by any computer hacker, and concerted sabotage could render a country unable to function."[83] Indeed, a U.S. intelligence official has claimed that $1 billion dollars and twenty capable hackers could "shut down America."[84] Thus the specter of information disruption can end up "weakening traditional hierarchical structures" and "eroding some traditional prerogatives of national sovereignty."[85]

One illustration of this potential superiority of offense over defense revolves around perhaps the most common means of attempting to defend against security vulnerability—the development of encryption technologies. Encryption attempts to protect sensitive information by coding it in such a way that only authorized individuals can determine what the data really mean. Traditionally used by intelligence agencies over the years, encryption has now spread to become commonplace in a wide array of government and financial service organizations. Unfortunately, encryption is a two-edged sword, and terrorist groups and transnational criminal organizations have become at least as adept at it as those institutions they attack, with a growth in their use of encryption estimated at between 50 and 100 percent a year: "Encryption compounds the problem of the original illicit activity by making it even more difficult to trace the proceeds of the crime or to unravel the records needed to investigate and prosecute the offender," and thus "perpetrators are able to operate with impunity because the significant human and financial resources needed to decode encrypted messages ensure that the investment will be made only for the most important cases."[86] Most targets of this disruptive use of encryption are unable to defend themselves against it, as the National Research Council recently argued that "current U.S. encryption policy is inadequate to meet the increased need for information security," and as a result the technique has been effectively used to deny access to records of deadly transfers of all sorts, including drug shipments and transnational terrorist activities.[87]

Finally, perhaps the broadest danger from information warfare is the undermining of a state's political authority. All systems of authority rely on command-and-control systems to function, depending on database management and effective communication among components as the means of keeping track of what is happening, making decisions, and providing necessary services. National defense systems, in particular, are heavily dependent on this kind of elaborate network to provide an early warning of any impending danger and to formulate quick and effective responses. In recent years, because of the technological breakthroughs in the computer industry, the vast majority of this information is stored electronically, and most of this communication is done electronically. But the very features that make this electronic authority structure so efficient, at the same time make it extremely penetrable. Maintaining authority in such circumstances necessitates not only safeguards against unwanted external intrusions into data and communications systems but also immediate detection and speedy restoration when such intrusions have occurred; it is in these second two areas that current defensive capabilities are especially woeful. Since legitimate authority depends heavily on data and communication credibility, the vulnerability is huge.

Conditional Utility of Information Warfare

Contemplating the use of information warfare or psychological operations raises some important moral and practical questions:

> Any discussion of information warfare, netwar, cyberwar, or even perception manipulation as a component of command and control warfare by the armed forces of the United States at the strategic level must occur in the context of the moral nature of communication in a pluralistic, secular, democratic society. That is, the question must be raised whether using the techniques of information warfare at the strategic level is compatible with American purposes and principles.
>
> Likewise, the question must be raised whether the armed forces of the United States have either the moral or legal authority and, more importantly, the practical ability to develop and deploy the techniques of information warfare at the strategic level in a prudent and practical manner. There are good reasons to be skeptical.[88]

Different cultural values or international predicaments would elicit different responses to these critical queries.

Most basically, there is some real uncertainty whether any critical computerized networked system can ever be made impervious to disruption through information warfare. Governments have spent large amounts of money and time trying to secure their systems against hacking and virus injection from the outside, even employing convicted former hackers as consultants, but these systems still are incredibly vulnerable. No security system that can persist safely over time has been discovered as yet.

Several specific conditions appear to be necessary for information warfare to function successfully. There is a striking parallel between the circumstances under which information disruption works best and the circumstances under which psychological operations work best, but nonetheless the conditional utility of each deserves separate treatment. As with the propositions concerning precision-guided munitions and nonlethal weaponry, the ideas presented here are inescapably tentative and preliminary.

Information disruption appears to work best when the initiator possesses an asymmetrical advantage in its ability to undertake techniques of information warfare, or superb countermeasures to reduce vulnerability, such that commencing its use will not result in retaliation far more intrusive than the initiating state expects. The U.S. Department of Defense, for example, has long recognized "the need to establish superiority in information warfare."[89] Because this instrument of bloodless war is more likely than precision-guided munitions or nonlethal weaponry to diffuse, and its initiation is thus more likely to result in retaliation of a similar kind, the best kinds of targets are those who lack the means to do so. Functionally this points to underdeveloped or technologically backward targets as optimal. In a related manner, information disruption may work best when the initiator can successfully overload the target's information or communication systems with noise, preventing it from functioning properly. This condition has two prerequisites: the initiator needs to have the electronic capacity to flood the target with extraneous messages, viruses, hacks, or the like in sufficient quantity to paralyze the target system, and the target needs to lack the capacity to filter out unwanted incoming data or to inoculate its systems against such intrusion. Even within advanced industrial societies, there are sectors that lack this kind of defense against external interference.

Second, information disruption seems to be most effective when the target has vast amounts of highly sensitive security information stored in electronic form and relied on for military command-and-control functions. Here the most advanced industrial societies are the most vulnerable, as their dependence on this kind of data—where almost everything vital seems to be highly digitized, automated, and mechanized—is the highest. The rationale for this claim is that electronic databases are the most vulnerable to unwanted intrusion, particularly (though not exclusively) if they are connected through a network, which is almost always the case today: it is clear that "increasing connectivity is a key enabler of cyber attack."[90] In a related manner, this vulnerability to information disruption rises when databases or communications systems are highly centralized—largely located in a few locations—rather than decentralized and spread in many locations. Often, such databases are centralized in many states, ironically, for security reasons, to focus one's scarce military protection on just a few spots, but this management system can easily backfire. The logic here is transparent—it is much easier to penetrate or disable or even demolish a single installation than multiple ones, and this would be particularly devastating if all the backups for vital security information were in one place.

Third, information disruption appears to function most smoothly when integrated fully into other military and political initiatives to disrupt the target. Specifically, information disruption options "seldom can 'stand alone,'" as they work best when combined with "other measures such as presence, force movements (e.g., movement into theater; call up of reserves), and other direct deterrent actions that serve as a demonstration of will."[91] Like the other instruments of bloodless war—precision-guided munitions and nonlethal weaponry—a widespread pattern is to implement information disruption initiatives in a manner totally divorced and disconnected from other coercive initiatives during a military confrontation. This reduces effectiveness because often the timing of the disruption is as important as the choice of which data and communications systems to target. In a parallel fashion, it is also important that "no single service or agency is or should be designated as responsible for IW [information warfare] actions in isolation from others; at the national level, IW strategy is needed and it must have full interagency involvement."[92]

We turn for comparison to psychological operations, which appear to work best against a target when it does not have easy access to other sources of information about what is actually transpiring and when other states are not trying to influence the target in contrary directions. The ideal target state is thus one that is completely isolated and subject only to the attacker's own propaganda, where the citizens cannot hear conflicting evidence or interpretations even from their own government. In today's world this scenario is not very common at all, but the least-developed societies in which people do not have access to television, radio, newspapers, or the Internet would be the closest approximation. If some competing sources of information are available, one's psychological operations have to be louder and slicker to be effective. Of course, even with a totally monopolized target, propaganda choices must be prudent—for example, "media manipulation that involves government personnel providing false information is neither politically wise nor consistent with U.S. policy and law."[93]

Second, psychological operations are likely to be most effective when an initiator can utilize easy-to-understand central images and slogans immediately intelligible to the target audience. For example, leaflets with complex, detailed messages do not work as well as those that have a few carefully chosen phrases and pictures to incite the recipients to anger or sympathy. It is clear that, in order to create such images and slogans, the initiator needs a comprehensive and sensitive understanding of the target culture and its members' preexisting concerns and perceptions. This level of understanding often requires years of background preparation and immersion and cannot be acquired quickly (or through crude monetary inducements), and historically the United States—whose citizens often have great difficulty learning even a single foreign language at the most rudimentary level—has great difficulty in this area. It is common to point out that among existing limits of psychological operations are the need for extensive planning and deployment time, lack of complete information from intelligence agencies, difficult coordination between military units and civilian information agencies, lack of qualified linguists, and poor understanding of cultural, political, economic, social, and ideological conditions.[94] Tuning a message for members of another society in such a way that it really touches a chord and sways their opinions is a real art. The key to success appears to be "to use credible messages and conduct surveys on the target audience's behaviors, attitudes and beliefs," and then "to effectively persuade citizens to change their political allegiances, lay down their arms and the like, the military must target their vulnerabilities and susceptibilities."[95] It is also clear that such psychological operations work best when targets believe that the sources from which they are used to receiving information are credible, and so a foreign initiator using this technique during a military confrontation would have to work very hard either to disguise itself as the source or to convince the recipients that its information is superior in accuracy to anything else available. Creating an image of source credibility is perhaps the most challenging goal in this regard, as much of the world's population has become unreceptive, cynical, or downright antagonistic toward the flood of crosscutting messages received each day.

Third, the state of mind of the target affects the success of psychological operations. On the one hand, this technique appears to work best against targets who are not zealous or ideologically committed, those who instead possess certain doubts or ambivalence about their cause even before receiving the propaganda. Such a situation fosters an openness or even curiosity toward outside information, creating a clear avenue through which one's own message can enter the picture. On the other hand, psychological operations seem to function better during a chaotic atmosphere of fear and panic, where people do not have the time to crosscheck information for accuracy and are jolted dramatically—often through the events of an ongoing military confrontation—out of their comfortable set of assumptions about whom to trust and how things operate. Indeed, propaganda tactics "can be extremely effective in wartime, when fear and emotions allow expert communicators to shape opinions and influence behavior on both political and military fronts."[96] So, combining the two ingredients, neither a perfectly rational nor perfect irrational target is ideal in this regard: the ultimate utility of psychological operations may thus revolve around international predicaments in which targets are either unable or unwilling to be certain about what is true and what is false.

Conclusion

Considered specifically in the context of the quest for bloodless war, this chapter has suggested some areas where offensive and defensive information warfare is most and least compatible with the value of casualty minimization and the broader goals of national security. It is clear that the specter of a completely interpenetrated and vulnerable world is a bleak one. Nobody's information would be safe, and nobody could be sure what to believe because there would be no reliable source through which to cross-check information. Military command-and-control would become haphazard at best, and it would be difficult to achieve decisive outcomes to wars.

Notes

1. This chapter draws heavily from Robert Mandel, *Deadly Transfers and the Global Playground: Transnational Security Threats in a Disorderly World* (Westport, CT: Praeger, 1999), chap. 9.

2. Martin C. Libicki, "Information War, Information Peace," *Journal of International Affairs* 51 (spring 1998): 411–412.

3. Ibid., pp. 416–417.

4. Joe Havely, "Why States Go to Cyber-War" (February 16, 2000), available online at http://news.bbc.co.uk/1/hi/sci/tech/642867.stm.

5. William B. Scott, "Information Warfare Policies Called Critical to National Security," *Aviation Week and Space Technology* 145 (October 28, 1996): 60.

6. "Clinton Girds U.S. for Terrorism War," *USA Today,* May 22, 1998, p. 1.

7. Gregory J. Rattray, *Strategic Warfare in Cyberspace* (Cambridge: MIT Press, 2001), p. 141.

8. Captain David Westover and Juan-Carlos Molleda, "A Brief History of U.S. International Radio Broadcasting and War: From the Voice of America to Radio Tikrit," (April 22, 2003), available online at www.psywarrior.com/westover.html.

9. Tom Regan, "Wars of the Future . . . Today," *Christian Science Monitor,* June 24, 1999, p. A1.

10. John Kerry, *The New War* (New York: Simon & Schuster, 1997), pp. 127–128.

11. Richard O. Hundley and Robert H. Anderson, "Emerging Challenge: Security and Safety in Cyberspace," in John Arquilla and David Ronfeldt, eds.,

In Athena's Camp: Preparing for Conflict in the Information Age (Santa Monica, CA: RAND, 1997), p. 231.

12. Libicki, "Information War, Information Peace," p. 417.

13. Rattray, *Strategic Warfare in Cyberspace,* pp. 99–100.

14. Brian C. Lewis, "Information Warfare," available online at www.fas.org/irp/eprint/snyder/infowarfare.htm.

15. Ibid.

16. Ibid.

17. Joint Publication 1–02, *Department of Defense Dictionary of Military and Associated Terms* (Washington, DC: U.S. Government Printing Office, March 23, 1994).

18. Melissa Dittmann, "Operation Hearts and Minds: Psychological Operations Are Becoming a Regular Part of Military Strategy," *Monitor on Psychology* 34 (June 2003): 32.

19. Bruce D. Berkowitz, "Warfare in the Information Age," *Issues in Science and Technology* 12 (fall 1995): 59–66.

20. Kerry, *New War,* p. 123.

21. Hundley and Anderson, "Emerging Challenge," p. 232.

22. Office of the Undersecretary of Defense for Acquisition and Technology, "Report of the Defense Science Board Task Force on Information Warfare-Defense" (November 1996), available online at http://jya.com/iwd.htm.

23. Owen E. Jensen, "Information Warfare: Principles of Third-Wave War," *Airpower Journal* 8 (winter 1994): 35–43.

24. Ibid.

25. George J. Stein, "Information War—Cyberwar—Netwar," in Barry R. Schneider and Lawrence E. Grinter, eds., *Battlefield of the Future: A Twenty-First Century Warfare Issue,* Studies in National Security no. 3 (Maxwell Air Force Base, AL: Air War College, September 1995), available online at http://www.airpower.maxwell.af.mil/airchronicles/battle/chp6.html.

26. Gary F. Wheatley and Richard E. Hayes, *Information Warfare and Deterrence* (Washington, DC: National Defense University Press, December 1996), chap. 2, available online at http://www.ndu.edu/inss/books/Books%20-%201996/Information%20Warfare%20and%20Deterence%20-%20Feb%2096/ch2.html.

27. Commander Randall G. Bowdish, "Information-Age Psychological Operations," *Military Review* (December 1998–January 1999): 31.

28. P. W. Singer, *Winning the War of Words: Information Warfare in Afghanistan,* Foreign Policy Studies Analysis Paper no. 5 (Washington, DC: Brookings Institution, October 23, 2001), available online at http://www.brookings.edu/views/articles/fellows/2001singer.htm.

29. Roger C. Molander and Sanyin Siang, "The Legitimization of Strategic Information Warfare: Ethical Considerations," *Professional Ethics Report* (American Academy for the Advancement of Science) 11 (fall 1998), available online at www.aaas.org/spp/dspp/sfrl/per/per15.htm.

30. Major Karl Kuschner, "Legal and Practical Constraints on Information Warfare" (1996), available online at www.airpower.maxwell.af.mil/airchronicles/cc/kuschner.html.

31. Jeff Glasser, "Psychological Operations: Getting the U.S. Message to Iraqis," *U.S. News & World Report* (March 22, 2003), available online at www.usnews.com/usnews/iraq/articles/qatar030321.htm.
32. Dittmann, "Operation Hearts and Minds," p. 32.
33. Bowdish, "Information-Age Psychological Operations," p. 28.
34. R. L. DiNardo and Daniel J. Hughes, "Some Cautionary Thoughts on Information Warfare," *Airpower Journal* (winter 1995), available online at www.airpower.maxwell.af.mil/airchronicles/apj/dinardo.html.
35. Dittmann, "Operation Hearts and Minds," p. 32.
36. CNN News, "Fierce Cyber War Predicted" (March 3, 2003), available online at www.cnn.com/2003/tech/ptech/03/03/sprj.irq.info.war.ap.
37. Bruce D. Berkowitz, "War Logs On: Girding America for Computer Combat," *Foreign Affairs* 79 (May–June 2000): 12.
38. Kuschner, "Legal and Practical Constraints on Information Warfare."
39. John Arquilla and David Ronfeldt, "Cyberwar Is Coming!" in Arquilla and Ronfeldt, *In Athena's Camp,* p. 23.
40. CNN News, "Fierce Cyber War Predicted."
41. John Arquilla and David Ronfeldt, "Information, Power, and Grand Strategy in Athena's Camp: Section 1," in Arquilla and Ronfeldt, *In Athena's Camp,* p. 141.
42. Jeffrey R. Cooper, *The Emerging Infosphere* (McLean, VA: Center for Information Strategy and Policy at Science Applications International Corporation, August 1997), p. 25.
43. John Arquilla and David Ronfeldt, "A New Epoch—and Spectrum—of Conflict," in Arquilla and Ronfeldt, *In Athena's Camp,* p. 5.
44. Rattray, *Strategic Warfare in Cyberspace,* p. 2.
45. Ken Silverstein, "Buck Rogers Rides Again: A 'Revolution' in High-Tech Systems Promises Big Profits and for the U.S. Risk-Free War," *Nation* 269 (October 25, 1999): 23.
46. Rattray, *Strategic Warfare in Cyberspace,* pp. 309–442.
47. CNN News, "Fierce Cyber War Predicted."
48. Westover and Molleda, "Brief History."
49. Kerry, *New War,* pp. 121, 124, 126, 128.
50. Rattray, *Strategic Warfare in Cyberspace,* p. 103; and Bradley Graham, "Eleven U.S. Military Computer Systems Breached This Month," *Washington Post,* February 26, 1998, p. A01.
51. Chris O'Malley, "Information Warriors of the 609th," *Popular Science* 251 (July 1997): 70–74.
52. Ibid.
53. Regan, "Wars of the Future," p. A3.
54. Charles Knight, Lutz Unterseher, and Carl Conetta, "Reflections on Information War, Casualty Aversion, and Military Research and Development After the Gulf War and the Demise of the Soviet Union" (1992), available online at www.comw.org/pda/0003refl.html.
55. Andrea Stone, "Soldiers Deploy on Mental Terrain," *USA Today,* October 3, 2001, p. 7A.
56. Singer, *Winning the War of Words.*

57. Westover and Molleda, "Brief History."
58. Colonel James W. McLendon, "Information Warfare: Impacts and Concerns," in Schneider and Grinter, *Battlefield of the Future*, chap. 7, available online at http://www.airpower.maxwell.af.mil/airchronicles/battle/chp7.html.
59. David A. Fulghum, "Telecom Links Provide Cyber-Attack Route," *Defense* 151 (November 8, 1999): 81.
60. David A. Fulghum and Robert Wall, "Combat-Proven Infowar Remains Underfunded," *Information Warfare* 154 (February 26, 2001): 53.
61. Westover and Molleda, "Brief History."
62. Jim Wolf, "U.S. Preparing for Cyber War," *Toronto Star*, November 9, 2001, p. E4.
63. Stone, "Soldiers Deploy on Mental Terrain," p. 7A.
64. Westover and Molleda, "Brief History."
65. BBC News, "US Escalates Psy-Ops War" (February 27, 2003), available online at http://news.bbc.co.uk/1/hi/world/middle_east/2805127.stm.
66. Thom Shanker and Eric Schmitt, "Firing Leaflets and Electrons, U.S. Wages Information War," *New York Times*, February 24, 2003, p. A1.
67. Westover and Molleda, "Brief History."
68. John R. MacArthur, "Thinking Big: The Propaganda Wars in the Psychological Struggle; Nations Wield Their Weapons of Mass Persuasion," *Boston Globe*, March 9, 2003: p. D12.
69. Shanker and Schmitt, "Firing Leaflets and Electrons" p. A1.
70. Anne Scott Tyson, "Hearts, Minds, Leaflets: War's Psychological Side," *Christian Science Monitor*, January 30, 2003, p. 2.
71. Max Boot, "The New American Way of War," *Foreign Affairs* 82 (July–August 2003): 53–54.
72. Shankar Vedantam, "Propaganda Seen As Key for Military, World Opinion; Leaflet Drops Are Only Part of U.S. Psychological Efforts," *Washington Post*, March 24, 2003, p. A18.
73. Westover and Molleda, "Brief History."
74. DiNardo and Hughes, "Some Cautionary Thoughts."
75. Lewis, "Information Warfare."
76. Havely, "Why States Go to Cyber-War."
77. Neil Munro, "The Pentagon's New Nightmare: An Electronic Pearl Harbor," *Washington Post*, July 16, 1995, p. C3.
78. DiNardo and Hughes, "Some Cautionary Thoughts."
79. Rattray, *Strategic Warfare in Cyberspace*, p. 141.
80. Louise I. Shelley, "Crime and Corruption in the Digital Age," *Journal of International Affairs* 51 (spring 1998): 605–611.
81. Rattray, *Strategic Warfare in Cyberspace*, pp. 17–20.
82. Berkowitz, "Warfare in the Information Age," pp. 59–66.
83. Walter Laqueur, "Postmodern Terrorism," *Foreign Affairs* 75 (September–October 1996): 35.
84. Ibid.
85. Brian Nichiporuk and Carl H. Builder, "Societal Implications," in Arquilla and Ronfeldt, *In Athena's Camp*, pp. 296–297.
86. Shelley, "Crime and Corruption in the Digital Age," pp. 613–615.

87. Ibid., pp. 613–614.
88. Stein, "Information War."
89. Lewis, "Information Warfare."
90. E. Anders Eriksson, "Information Warfare: Hype or Reality?" *Nonproliferation Review* 6 (spring–summer 1999): 61.
91. Wheatley and Hayes, *Information Warfare and Deterrence*, chap. 3.
92. Ibid.
93. Ibid.
94. Bowdish, "Information-Age Psychological Operations," p. 31.
95. Dittmann, "Operation Hearts and Minds," p. 32.
96. Vedantam, "Propaganda," p. A18.

6

What Can Go Wrong

Even for many of those who believe that the quest for bloodless war is a noble pursuit, major areas of concern emerge about its application in practice. Are there dangers—specifically in terms of negative repercussions or possible backfire effects—that can occur if casualty aversion strategies are the primary means for conducting military combat operations overseas? A wide range of drawbacks exist, most of which emerge as a consequence of casualty aversion having too great an influence on the nature of foreign military action. Although earlier chapters highlighted the special dangers surrounding precision-guided munitions, nonlethal weaponry, and information warfare, this chapter analyzes the overarching concerns, summarized in Figure 6.1, in a broader context.

Most generally, some onlookers suggest that the quest for bloodless war may undercut the very foundation of national security:

> Postulating the employment of remote lethal targeting technology to wage war—followed by the unopposed deployment of peacekeepers—has given rise to the hopeful but misplaced belief that future wars can be fought with little or no loss of American lives. . . . The military can't protect both Americans and innocent populations abroad by adopting zero-casualty force protection as an operational priority. There are causes for which our soldiers should be willing to fight and die.[1]

The basic crippling of military power, compounded by the distortion of mission priorities, lies at the heart of this worry:

Surely we must make a distinction between, on the one hand, the moral and political imperative of shielding military forces from risks that are superfluous to the accomplishment of operational and strategic objectives and, on the other hand, the subordination of those objectives to pursuit of the ideal of bringing every soldier home alive. Casualty-phobic timidity on the battlefield can be just as self-defeating as bloodthirsty recklessness.²

Thus casualty aversion can undermine fundamental security objectives.

Within the national security area, the span of skepticism about possible negative effects from casualty minimization is truly vast, encompassing such fears as onlookers' horror and censure at any loss of life, initiator ability to engage in unchecked, cost-free foreign military action and external domination, widespread obliviousness to ongoing warfare due to its lack of effect on the functioning of proximate societies, initiator lack of motivation or ability to achieve decisively stable and successful outcomes, accelerated enemy development of counterstrategies, and overall escalation of the monetary costs of warfare. Many of these dangers constitute prime targets for those who are fundamentally opposed to casualty aversion. Although it would be impossible to estimate the exact probability of occurrence of each source of concern, an analysis of the primary worries—involving both political policy initiation and military operational style—seems critical to isolating the value of the quest for bloodless war.

Figure 6.1 Dangers of the Quest for Bloodless War

Dismal International Relations
Unrealistic external yardsticks: overly optimistic onlooker expectations
Loss of global credibility: image of weakness among friends and foes alike

Increased Vulnerability to Disruption
Evening of the playing field: inadvertent technology diffusion
Cycle of insecurity-promoting competition: offensive-defensive technology race

Crippled Military Might
Limited palette of military action: narrowing/skewing the range of viable military options
Crisis of confidence in the military: erosion of the military ethos

Endless Imprudent Conflicts
Unrestrained political initiation of war: foreign policy recklessness
Longer and more frequent wars: indecisive military action

Overly Optimistic Onlooker Expectations

Examining the dangers more specifically, unrealistic onlooker expectations about unintended carnage represent a key concern. The overly optimistic outlook alluded to here foresees both reduced destruction and reduced disruptive impact:

> The number of men involved in arms factories and armies will decline precipitously—one projectile can be fired for every thousand previously needed. More important, the level of devastation will decline as well. . . . More precisely, in seeing the end of total war, we see an end to an era where war puts society's very being at stake. Regimes may rise and fall, but as in the premodern era, the life of ordinary men will go on.[3]

The common underlying assumption here is that "we know exactly what we are doing, we almost never miss . . . and we have virtually no casualties."[4]

These expectations about the benefits of casualty aversion may be way too high in terms of protection of human life:

> Culturally, American society has a greater distrust of government than many other societies, a sentiment which has historic roots, but which has deepened since Vietnam and Watergate. This distrust means that deaths "for state purposes" must be shown to be necessary, purposeful, and unavoidable. Together with the belief, encouraged by military and political leaders, that war can be—indeed should be—free of American casualties, this distrust can lead some to argue that any such deaths are the result of government incompetence or deceit.[5]

The quest for bloodless war may not result in as much protection of civilians as one might initially anticipate: "By means of 'precision' bombardment or cyber-weapons, such attacks focus on centers of political authority and sources of national strength (including industry, communications, and economic infrastructure)"; as a result, "the impact (both short-term and long) on civilian populations of strategic attack can be profound—even when extraordinary measures are taken to limit immediate collateral damage."[6] In other words, the kind of damage necessary to hurt enemy leadership may inexorably seep down to hindering the lives of average citizens within affected societies.

When, as is inevitable, some unintended human and property damage occurs, the glowing rhetoric often accompanying casualty aversion can make the ensuing resentment and anger ever more fierce. The interpretation of onlookers may be that the accidental destruction of innocents must represent willful negligence and contempt for human life. Despite tangible increases in both efficiency and human protection, even the smallest mistake by an initiator—due to human or mechanical error—can produce this kind of scorn.

Image of Weakness Among Friends and Foes Alike

A state pursuing bloodless war could often convey an image of weakness to its enemies:

> The perception among our enemies and allies alike is that the American public is unwilling to commit to any military operation in which one can expect even a minimal number of casualties. Furthermore, they believe that once an enemy engages the United States, it can force the latter to withdraw from its commitments when American casualties mount. Because of our casualty aversion, in the eyes of the world, we are becoming "a sawdust superpower."[7]

It is terribly difficult to maintain credibility as a global superpower without a widely acknowledged will—as well as capacity—to use lethal force when necessary to achieve stated objectives. An overemphasis on casualty aversion can prevent military commanders from undertaking bold but relatively risky initiatives that have the potential to save lives or shorten a conflict in the long run.

Seeing vulnerability, savvy adversaries can often easily exploit it to their advantage:

> Moreover, our opponents know our limitations and work to take advantage of them. Saddam speculated openly about our willingness to see the body bags come home. Aideed found that one very bad night was enough to start our planning for a withdrawal from Somalia. He also had his forces use women and children as shields in encounters with US soldiers, believing this would unnerve them—which it did. Such experience then becomes data for the next thug.[8]

An adversary devoid of the constraints of casualty minimization can thus feel emboldened. Recent conflicts have substantiated this weakness: "Lots of people around the world believed casualty phobia was the Achilles' heel of American foreign policy, and some even acted accordingly: Slobodan

Milosevic, Osama bin Laden, and now Saddam Hussein."[9] The 2003 war against Iraq vividly exemplified this vulnerability, and Saddam Hussein probably at least in the early stages of the war drew strength from the West's casualty aversion because "he could mistake the U.S.-British anxiety about body bags as a lack of resolve."[10] However, it is only fair to note that the entrenched anti-Americanism in much of the third world may prevent ungarbled interpretation of esoteric signaling by the United States through its weapon choices.

Although attempting to protect human life is admirable, the zeal behind this end can easily boomerang in terms of enemy reactions:

> The impulse to avoid indiscriminately sacrificing the lives of American servicemen and women is noble, and it is certainly preferable to the callous attitude of the Russian military toward its tank and file, whether in Chechnya or aboard the *Kursk,* the submarine that sank in 2000. But in the immediate post-Cold War era America's political leaders were more careful to husband the lives of volunteers—men and women who have sought out combat and its risks—than their predecessors were with the lives of draftees in years past.
>
> Sometimes this policy can backfire cruelly: if foreign enemies know that killing a few Americans will drive the U.S. out of their country, they are far more likely to target American soldiers or civilians. And their task will be facilitated by force protection strategies—as at Khobar Towers in 1996 or in Beirut in 1984—that conveniently group American service members in one building. Every time U.S. forces flee some country after suffering casualties, it makes it less likely that the U.S. will be able to accomplish its objectives in the future *without* using force.[11]

Clever enemies know exactly how to take advantage of this kind of sensitivity.

The underlying logic behind this cynical enemy interpretation of wartime casualty aversion relies on an asymmetrical calculus of how to defy and defeat a superior military power, clarifying the dangers of acting "as if the public is casualty phobic when the public is not." Among recent U.S. adversaries, for example, "the key was to defeat the U.S. will by raising the costs of war beyond what the American public was willing to pay—something each of the challengers thought was within his grasp."[12] More specifically, would-be foes can exploit the social dimension of strategy by figuring that, "if you can make the war long, if you can make the war bloody, perhaps even not with the blood of Americans but with the blood of civilians on the other side," then "you can offset the American advantage in technology."[13] Regardless of the accuracy of this external perception, it can be truly debilitating for effective foreign security policy in general and for deterrence strategies in particular.

In a parallel manner, even allies may frequently draw negative conclusions from an exhibition of casualty aversion. Many analysts believe that "predicating American capabilities and policy upon the quick, costless war presumption will inevitably call into question the credibility of the American commitment to its allies, thus making the formation of multilateral military initiatives considerably more difficult—why should our allies commit themselves to operations that we are not willing to undertake alongside them?"[14] It can become highly unpredictable to depend on a military alliance with a state that will flee at the sight of body bags. In a situation where allies perceive the United States specifically as being casualty-averse, and a need arises for more troops from coalition partners to go into a combat zone, these allied states may be understandably reluctant to do so. Moreover, this reluctance can dramatically reduce the value of allied troops as a deterrent to prevent enemies from launching initiatives. This reluctance can be accentuated if the United States wishes to fly low-casualty combat sorties from the air while expecting allies to undertake the more casualty-intensive "dirty work" on the ground. Of course, limited defense budgets and interoperability problems already severely limit allies' abilities to fight alongside the United States.

Inadvertent Technology Diffusion

Western adoption of technologies useful for casualty aversion may under very limited circumstances risk their spread to hostile states and terrorist organizations. Globalization and porous national boundaries make it relatively difficult to keep any weapons technology proprietary for very long in today's world. Moreover, casualty-minimizing technologies are likely to become less expensive, and there is a long tradition of emulating arms that are small, cheap, and successful: "Improved information technology produced for solely commercial ends will allow more people access to more lethal and more precise weapons at a lower cost than is imaginable today."[15] Some of these technologies, even though they minimize loss of human life, may prove to be advantageous for enemies to cripple advanced industrial societies. Unsavory transnational groups or rogue states could use any one of an assortment of these technologies to incapacitate the U.S. government or business sectors with relative ease. Even though the most advanced precision-guided munitions, nonlethal weaponry, and information warfare technologies might be difficult for novices who stole them to utilize in combat, this may not prove to be a permanently insurmountable obstacle. These disruptive efforts could be difficult for any victim to stop due to the illicitly obtained state-of-the-art technology employed. Of course, the United States is not as vulnerable as many other Western states would be to the proliferation of these weapons technologies due to its relatively insulated geopolitical position and its huge military power advantages.

Adversaries could utilize diffused technologies associated with casualty aversion not only against other states in a military confrontation but also against their own people. For example, precision-guided munitions, nonlethal weaponry, and information warfare provide potentially superb tools for brutally suppressing an otherwise less-than-supportive domestic population (in fairness, of course, traditional casualty-maximizing arms such as battle tanks, poison gas, and antipersonnel mines would be at least as problematic when used in this manner). Although the West (from which the majority of technologies useful for casualty aversion emerge) would not necessarily witness international instability directly resulting from the uncontrolled proliferation of these technologies, it would very well witness an upsurge in internal human rights violations for which it would feel at least partially responsible. In such cases, the principle of self-determination of peoples may prevent outraged outside powers from readily intervening to put a stop to these abuses.

Offensive-Defensive Technology Race

The spread of technologies useful for casualty aversion could, however, indirectly trigger a destabilizing new arena of international technology competition. This race could drain national budgets and reduce the exploration of other types of useful military technology. Even though many casualty avoidance technologies are designed primarily to disable an adversary's ability to respond, the widespread use of these technologies could accelerate the sophistication of techniques that could be used so rapidly that offense would outstrip defense, meaning that no state would be able to protect itself from intrusion from the outside. For example, current defenses against precision-guided munitions, nonlethal weaponry, and information warfare possess some glaring gaps and weaknesses that will need to be addressed if these technologies continue to proliferate.

A race would also likely occur to develop expensive countermeasures against these technologies useful for casualty aversion. Enemies of the United States "can be expected to develop countermeasures to the Pentagon's high-tech weapons—'that's part of the game'—but most will have a hard time competing; and earlier military revolutions show the importance of getting a jump on one's adversaries."[16] As the casualty aversion arsenal expands, threatened states will be driven—like it or not—to acquire protective or countermeasures to these technologies, including the "hardening" of potential targets, to block the opponent's efforts to disrupt and degrade their own capabilities. Thus the proliferation of the means for pursuing bloodless war may end up requiring the development of sophisticated new defenses on the part of those very societies who invented technologies useful for casualty aversion, societies who in some way are the most vulnerable to them because of the value placed on the well-being of soldiers and civilians.

Narrowing/Skewing the Range of Viable Military Options

Casualty aversion may serve to constrict the range of military options available. Specifically, it is possible that "the perception among civilian elites—the policy makers who determine national strategy—that the public is casualty-averse hinders coercive diplomacy and limits military options in support of our national strategy"[17] (although shrewd elites' use of an adversary's perception of their casualty aversion to enhance coercive diplomacy by undertaking a large-scale deployment could convey absolute credibility). The narrowed range of viable military options during wartime may include "restrictions on where troops can go, which areas receive aid, and the types of military activities conducted."[18]

Because of the reduced palette of strategic choice during a military confrontation, the options ultimately selected may be less effective:

> Force-protection fetishism encourages military half-measures directed against symptoms rather than sources of international political instability. This was as true of the Gulf War as it was of Allied Force. In

both cases, the national leadership was not prepared to run the political and military risks necessary to achieve a strategically conclusive victory. Caution may well have been justified, but the chief consequence in the Gulf and the Balkans was the survival of two rogue regimes, one of them bent on massive revenge.[19]

Indeed, an "absolute dependence on high tech in pursuit of a bloodless war" may impede war-fighting doctrine because "it will limit the ability to respond to the full range of possible conflicts," increasing the chances of "communicating to potential enemies that the direct employment of ground combat troops in favor of other options is foreclosed."[20]

Key consequences of this constrained choice may include not only ineffective or inefficient execution of military operations but also self-censoring that causes leaders to unnecessarily refrain from making profitable use of the armed forces in pursuit of national objectives.[21] Given the ability to launch technologies useful for casualty aversion from virtually any setting at a wide range of distances, the result of a force protection focus could easily be heavy reliance in future wars on remote coercive efforts (associated especially with both precision-guided munitions and information warfare), at the neglect of the vital wartime role of ground soldiers fighting in close combat. Strategically, casualty phobia can foster encouragement of politically inconclusive coercive initiatives, as "casualty phobia invites half-baked uses of force."[22] A stark example of this limitation, where decisionmakers must weigh the importance of the target against the predicted noncombatant casualties, occurred during Operation Iraqi Freedom in 2003, when "some targets of military value were not struck because of the danger to civilians."[23]

An overemphasis on casualty aversion could also cause otherwise unwarranted support for certain weapons programs:

> A typical example came to light in the recent decision to end the Crusader artillery program. Informed that the program was under consideration for cancellation, Army officials attempted to defend the system from the Secretary of Defense's effort. Army talking points provided for lobbying members of congress suggested that the Secretary of Defense's cancellation of the system would put soldiers lives "at risk." This point was particularly inflammatory and seems to have had a lot to do with the rather nasty and public manner in which the issue was resolved.[24]

The quest for bloodless war can distort consideration of technology and manpower trade-offs that influence future force structure design and military acquisition programs.²⁵ Skeptics argue that today the Pentagon and weapons manufacturers frequently convey the impression that "all modern wars can be fought quickly, cleanly and with very few US casualties" to promote acquisition of "scores of costly high-tech weaponry."²⁶ In this way, it is possible that unscrupulous advocates within the military establishment could artificially leverage the issue of potential harm to soldiers in order to justify armaments expenditures, escalating the monetary costs of warfare while simultaneously de-escalating the human costs.

Erosion of the Military Ethos

From the military's own standpoint, perhaps the most disturbing prospect emerging from the quest for bloodless war is the potential erosion of the military ethos: the military ethic "is built on the principles of self-sacrifice and mission accomplishment; troops are supposed to be willing to die so that civilians do not have to."²⁷ This "warrior code" clearly encompasses why soldiers fight, how they fight, what brings them honor, and what brings them shame.²⁸ Many active-duty officers and military analysts alike see casualty aversion as a "disturbing trend"— "not only would it jeopardize the success of military missions," but also it would "corrode the military ethos of self-sacrifice and protection of noncombatants."²⁹ Soldiers may hesitate to undertake heroic acts to save brothers in arms under fire. Possible negative consequences from the transforming military ethos include "the degrading impact of casualty aversion in excessive force protection, which shifts mission risk from the US military to others,"³⁰ and an altered self-conception of soldiers emphasizing more ability to utilize sophisticated technology rather than exhibition of courage on the battlefield. Of course, the warrior code has varied considerably across time and culture,³¹ and so it would be unfair to judge the impact of the quest for bloodless war on it using a static and rigid yardstick.

In recent years specifically within the United States, a concern has emerged that the "'warrior culture' has declined, eroding Americans' willingness to tolerate the sacrifices necessary to sustain international leadership."³² It may be that "an overt stress on casualty limitation causes an army's leaders to become too timid, and the soldiers to lose their 'warrior spirit'":³³

> A possible by-product and serious danger of this heightened awareness [of enemy civilian and military casualties] is the fear that casualty-sensitivity may breed timidity in U.S. military leaders. This cascading effect was passionately echoed by Chairman of the Joint Chiefs of Staff (CJCS) General John Shalikashvili before the Senate Armed Services Committee. "I do not want to see us evolve to a point where we have expectations in this country of a war where nobody gets killed on our side, and where we don't have any collateral damage on the other side."[34]

Although "the current belief that Americans lack the requisite fortitude to endure prolonged military operations is ironic, because their history has been defined by success in such conflicts,"[35] this represents a genuine and widespread fear. However, "the effect of casualties on a military's psyche will not depend merely on the absolute losses suffered, but instead on how those losses were perceived by the body politic, which is, in turn, a function of the level of popular support for the war."[36]

It is worth noting, however, that the idealized motives and attitudes of soldiers themselves embody numerous apparent contradictions. They fervently hope for peace, more intently than many policymakers, yet they fight with passion and zeal in battle. They seek prudent force protection, but yet are willing to die for their cause. They are resolute and doggedly determined in their pursuit of their goals, yet show flexibility and responsiveness in their battlefield strategy and tactics. They dutifully and without question obey their political leaders, yet exercise independent command leadership on the field.

Foreign Policy Recklessness

The quest for bloodless war can also on occasion lead to foreign policy recklessness by political elites:

> The very technology that makes "bloodless" war possible may also serve to encourage the use of force in circumstances where perceptions of stakes and risks might otherwise counsel restraint. Is the United States in fact transforming war into essentially a riskless enterprise—i.e., one in which the level of risk is dwarfed by the magnitude and high probability of strategic payoff? If so, then what is to keep future Presidents from taking a casual approach to military intervention?

> Should the United States really look forward to creating a capacity to wage "painless" war, war without American death, war dangerous and hurtful only to the other side? Would not the prospect of casualty-free combat invite the use of force over even trivial stakes?[37]

Thus low-cost success could engender a kind of "throw-caution-to-the-wind" mentality in decisions about whether or when to intervene abroad, devoid of a sense of judicious restraint. For a state with aspirations of global hegemony, like the United States, this expectation of costless war could be particularly dangerous, even when tempered with its pronounced tendency of late to prefer brief and limited foreign military engagements. Ironically, then, at the same time wars or interventions stressing casualty minimization could signify to some a lack of resolve and willingness to take risks, these military initiatives could—in a rather contrasting fashion—reflect a lack of premeditated calculated caution about the true costs of war.

Casualty aversion strategies could also foster obliviousness to ongoing warfare due to its lack of effect on the functioning of proximate societies. For example, the development of both precision and nonlethal weaponry could induce the United States to abandon the notion that it should commit its forces to combat only as a last resort.[38] Thus casualty-minimizing warfare could emerge as a less horrifying and more routine and politically acceptable instrument of foreign security policy.

In a parallel manner, because of the presumed aversion to human sacrifice, casualty aversion can cause leaders to exaggerate the justifications for entering a war to overcome this source of queasiness:

> When state survival is not threatened . . . there must be a moral cause worthy of a crusade to justify a distant war for Americans. Not surprisingly, political leaders when seeking the ultimate sacrifice from Americans have had to exaggerate the stakes. These exaggerations often come back to trap them, forcing greater commitment than reality requires. We were told that freedom itself was at risk in Vietnam, that Noriega was a drug lord, that Saddam was a Hitler, that Aideed was a warlord, and that the Haitian rulers were brutality incarnate.[39]

Facilitating this tendency to overstate the stakes involved in a war are the ambiguities caused by casualty aversion in calculating meaningful coercive force ratios: for confrontations where national survival is not at stake, the very limited force usable by a casualty-minimizing state may be substantially less than its actual overall military strength.

Indecisive Military Action

The quest for bloodless war can directly backfire in terms of its dual aspirations to protect human life and to achieve decisive military victory. One dire outcome is that "the political instinct to avoid civilian casualties in the short term may result not only in unwise military strategies, but also greater civilian suffering."[40] In other words, efforts to minimize casualties in the short run can sometimes maximize casualties in the long run due to prolonged indecisive military action:

> Danger resulting from a misguided belief in bloodless conflict comes from turning abstract notions of battlefield fairness or proportionality into an operational imperative. America has a strong sense of fair play and justice for all. It abhors human suffering, a virtue which is among its greatest strengths. However, blindly applying fairness and balance on the battlefield is inimical to national security. History suggests that the denial of military experience increases the longterm suffering inherent in combat.
>
> Any military that limits itself to narrowly calibrated proportional force is an organization in search of defeat. The Armed Forces do not go to war to put up a good fight; they go to win. They do not attack in kind; they attack with every type of force to break enemy will and defeat it. By prosecuting warfare aggressively, one not only limits losses but shortens the conflict and thus lessens the suffering of noncombatants and often enemy forces themselves.[41]

This problem is exacerbated by the possibility that—despite the exaggerated justifications for entering war discussed earlier—sophisticated technologies useful for casualty aversion could make the conduct of war more unintelligible or confusing to the civilian citizenry, reducing its ability to participate effectively through the democratic process in decisions about engaging in and disengaging from foreign military confrontations.

The sense of personal and national vulnerability could also rise, as individuals will come to realize that—even with sizable military forces—there is no way really to protect oneself and one's society from the unorthodox methods of disruption advanced by technologies associated with precision-guided munitions, nonlethal weaponry, and information warfare. For the public in powerful countries like the United States, this recognition that large military forces and matériel do not secure safety from many kinds of international threats can be a painful one indeed. Because of the difficulties of deterring the use of such nonstandard coercive instruments, heightened emotional anguish among many of the citizens of advanced industrial societies seems virtually inevitable.

A fundamental question remains whether relatively bloodless armed conflicts, with little loss of life, injuries, or property damage among allies, neighboring populations, and even foes, can achieve decisively successful and stable outcomes. From a military standpoint, of course, there is no substitute for victory. Despite the changes in the type of warfare over time, this long history of bloody conflicts has a lot to tell us about the difficulties of obtaining stable victory without sacrifice. Obviously this combination is conceivable through, for example, surgical decapitation of target leadership or destruction of key industrial nodes; however, there is certainly no guarantee this will occur in the kinds of confrontations encountered today. Sometimes decisive outcomes still depend on what they used to—defeating ground troops in bloody battles.

During recent foreign military actions, casualty incidence has often shifted from the actual combat phase to the postcombat phase. In other words, a carefully delineated military campaign may be conducted with the appearance of success with few casualties, but then—after the large-scale fighting is over—snipers, saboteurs, guerrillas, and suicide bombers may persist in violent initiatives that cause more bloodshed after a war ends than during a war. Ironically, the application of casualty aversion strategies during warfare may leave more aggrieved and frustrated enemies around afterward to wreak havoc. No case illustrates this danger better than the 2003 war against Iraq, in which Americans have lost far more lives after the war than during it. Unfortunately, most discussion of casualty aversion has ignored the aftermath of war and this postconflict phase.

Conclusion

Perhaps the most important and all-encompassing danger is that, by diminishing the horrors of violent conflict, the quest for bloodless war may reduce both the negative incentives to initiate or perpetuate violence and the positive incentives to end a conflict. Carl von Clausewitz once remarked, "Let us not hear of generals who conquer without bloodshed, [for] if a bloody slaughter is a horrible sight, then that is a ground for paying more respect to war."[42] Even more succinctly, General Robert E. Lee in a similar vein claimed, "It is well war is so terrible, or we should grow too fond of it."[43]

The lesson from this discussion of dangers is not only that the emphasis on casualty aversion ought to be moderated in the context of other important political-military security values but also that the quest for bloodless war ought not to be applied indiscriminately regardless of the particular circumstances at hand. It is clear that even if minimizing bloodshed were the top priority in warfare, this would not mean that all military preparation and strategic planning and choice of coercive instruments would be guided exclusively by that value. And from this analysis it is not at all evident that the quest for bloodless war ought to be the top priority, or even one of the very top priorities, in some types of military confrontation. Protecting people—both civilians and combatants—during violent military confrontations is obviously very important, for both moral and practical reasons, but such an aspiration should not totally control or completely skew the nature of a military campaign.

Notes

1. Barry R. McCaffrey, "Lessons of Desert Storm," *Joint Forces Quarterly* (winter 2000–2001): 17.

2. Jeffrey Record, "Force-Protection Fetishism: Sources, Consequences, and (?) Solutions," *Aerospace Power Journal* (summer 2000): 5.

3. George Friedman and Meredith Friedman, *The Future of War: Power, Technology, and American World Dominance in the Twenty-First Century* (New York: St. Martin's Griffin, 1996), p. 393.

4. Anthony H. Cordesman, "The U.S. Bombing and Ramadan: The Real Problems We Face Because of Our Failure to Understand Asymmetric Warfare and Mistakes That Turn 'Information Dominance' into a Self-Inflicted Wound" (November 9, 2001), available online at www.csis.org/features/cord_011109.htm.

5. Harvey M. Sapolsky and Jeremy Shapiro, "Casualties, Technology, and America's Future Wars," *Parameters* (summer 1996): 123.

6. Carl Conetta, "The 'New Warfare' and the New American Calculus of War," (Cambridge, MA: Project on Defense Alternatives, Commonwealth Institute, September 30, 2002), available online at www.comw.org/pda/0209newwar.html.

7. Major Charles K. Hyde, "Casualty Aversion: Implications for Policy Makers and Senior Military Officers," *Aerospace Power Journal* 14 (summer 2000): 18; and Mark Conversino, "Sawdust Superpower: Perceptions of U.S. Casualty Tolerance in the Post–Gulf War Era," *Strategic Review* 25 (winter 1997): 22.

8. Sapolsky and Shapiro, "Casualties, Technology, and America's Future Wars, " p. 123.

9. Peter D. Feaver, "Casualties Are the First Truth of War—And One the Public Is Well Prepared to Accept," *Weekly Standard* 8 (April 7, 2003): 17–18.

10. John Chalmers, "Casualty Aversion Is Saddam's Greatest Weapon" (March 30, 2003), available online at http://uk.news.yahoo.com/030330/80/dwlff.html.

11. Max Boot, *The Savage Wars of Peace: Small Wars and the Rise of American Power* (New York: Basic Books, 2002), pp. 327–328.

12. Peter D. Feaver and Christopher Gelpi, *Choosing Your Battles: American Civil-Military Relations and the Use of Force* (Princeton: Princeton University Press, 2004), pp. 4, 209.

13. Admiral Gene LaRocque, "Approaching the Digital Battlefield" (Washington, DC: Center for Defense Information, December 15, 1996), available online at www.cdi.org/adm/transcripts/1014.

14. Andrew P. N. Erdmann, "The U.S. Presumption of Quick, Costless Wars," *Orbis* 43 (summer 1999): 379–380.

15. Paul Musgrave, "Will Technology Be Used to Make War More Humane? Warfare and the Information Revolution" (February 2003), available online at www.paulmusgrave.net.

16. Ken Silverstein, "Buck Rogers Rides Again: A 'Revolution' in High-Tech Systems Promises Big Profits and for the U.S. Risk-Free War," *Nation* 269 (October 25, 1999): 23.

17. Hyde, "Casualty Aversion," pp. 23–24.

18. Daniel Byman and Matthew Waxman, *The Dynamics of Coercion: American Foreign Policy and the Limits of Military Might* (New York: Cambridge University Press, 2002), pp. 184–185.

19. Record, "Force-Protection Fetishism," p. 8.

20. McCaffrey, "Lessons of Desert Storm," p. 17.

21. Richard A. Lacquement Jr., "Understanding the Casualty Aversion Assertion: Implications and Applications," paper prepared for presentation at the annual meeting of the International Studies Association, Portland, OR, February 26–March 1, 2003, p. 6.

22. Jeffrey Record, "Collapsed Countries, Casualty Dread, and the New American Way of War," *Parameters* (summer 2002): 14.

23. John T. Correll, "Casualties," *Air Force Magazine* 86 (June 2003): 52.

24. Lacquement, "Understanding the Casualty Aversion Assertion," p. 12.

25. Ibid., p. 6.

26. LaRocque, "Approaching the Digital Battlefield."

27. Peter D. Feaver and Christopher Gelpi, "A Look at Casualty Aversion: How Many Deaths Are Acceptable? A Surprising Answer," *Washington Post*, November 7, 1999, p. B3.

28. Shannon French, *The Code of the Warrior: Exploring Warrior Values Past and Present* (Lanham, MD: Rowman & Littlefield, 2003), pp. 1–18.

29. Tom Bowman, "War Casualties Could Test Public's Resolve: Officials Fear Support could Shrink As Troops Search for Bin Laden," *Baltimore Sun*, November 18, 2001, p. 19A.

30. Hyde, "Casualty Aversion," p. 25.

31. French, *Code of the Warrior*, pp. 21–29.

32. Andrew P. N. Erdmann, "The U.S. Presumption of Quick, Costless Wars," *Orbis* 43 (summer 1999): 373.

33. Karl W. Eikenberry, "Take No Casualties," *Parameters* (summer 1996): 116.

34. John N. Sims Jr., "Shackled by Perceptions: America's Desire for Bloodless Intervention," thesis presented to the faculty of the School of Advanced Airpower Studies, Maxwell Air Force Base, AL, June 1997, p. 3.

35. Erdmann, "U.S. Presumption of Quick, Costless Wars," p. 363.

36. Eikenberry, "Take No Casualties," p. 112.

37. Record, "Collapsed Countries," p. 20.

38. Karl P. Mueller, "Politics, Death, and Morality in US Foreign Policy," *Aerospace Power Journal* (summer 2000): 15.

39. Sapolsky and Shapiro, "Casualties, Technology, and America's Future Wars," p. 124.

40. Project on the Means of Intervention, "Understanding Collateral Damage" (Washington, DC: Harvard University, John F. Kennedy School of Government, Carr Center for Human Rights Policy, June 4–5, 2002), p. 5.

41. McCaffrey, "Lessons of Desert Storm," p. 17.

42. Carl von Clausewitz, *On War,* ed. Anatol Rapoport (Baltimore: Penguin Books, 1968), p. 345.

43. Douglas Southall Freeman, *R. E. Lee: A Biography,* vol. 2 (New York: Charles Scribner's Sons, 1941), p. 462.

7
Toward Effective Casualty Aversion

Due to the dangers associated with the quest for bloodless war, there is a need for conditional analysis of the utility of casualty aversion strategies. While this book questions those who view bloodless war as a panacea and universal guiding principle for the future, there is no intent here to imply that casualty aversion is never useful. The conditions mentioned in this chapter, summarized in Figure 7.1, are overarching ones intended to cover broadly all the means of casualty aversion.

Nonetheless, significant differences exist—as earlier chapters have indicated—among the circumstances promoting the optimal utility for precision-guided munitions, nonlethal weaponry, and information warfare. Several types of nonlethal weaponry are well suited for crowd control and for use against unruly subnational groups, while precision-guided munitions are not. Similarly, nonlethal weaponry is more useful to address small acts of spontaneous disruption, while precision-guided munitions and information warfare seem more attuned to large-scale premeditated and organized coercive operations. The application of nonlethal weaponry works better in combination with economic sanctions than is the case with either precision-guided munitions or information warfare. Information warfare has special advantages in confronting electronic command-and-control systems, while nonlethal weaponry, primarily designed to incapacitate humans, generally does not. Information warfare and precision-guided munitions work best when enemy capabilities are highly centralized, while such is not the case for nonlethal weaponry. Finally, information warfare works best when a target is relatively isolated, which is not necessarily the case for either precision-guided munitions or nonlethal weaponry.

Figure 7.1 Conditional Utility of the Quest for Bloodless War: Conditions Most Favorable to Effective Casualty Aversion

Legitimate Peripheral Interests of Initiators
Remote but reasoned national involvement in a conflict
Carefully specified and articulated military objectives

Vulnerable Character of Targets
Rational state targets without passionate resolve for their cause
Existence of pivotal fixed identifiable targets

Discrete and Restricted Nature of Missions
Short-term missions (especially humanitarian ones) with limited objectives
Attempts to achieve influence over a target that can be rehabilitated

Capacity to Discriminate During Combat
Outstanding electronic and human intelligence capabilities
Easy ability to differentiate civilians from combatants

Superiority of Advanced Casualty Aversion Technology
Possession of asymmetrical advantage in casualty aversion technology
Unambiguous presence and intention to use casualty aversion technology

Isolated and Integrated Thrust of Action
Existence of containable, localized conflict devoid of contagion
Integration of casualty aversion with other military strategies

One omnipresent background issue affecting the utility of casualty aversion strategies is the possibility that mechanical failure and human error can be rather unpredictable sources of interference. The inherent fallibility of all weapons and of the humans utilizing them prevents universal effectiveness in war. The special complexity and sophistication of the casualty-minimizing technologies—precision-guided munitions, nonlethal weaponry, and information warfare—make them especially susceptible to these weaknesses. Extensive training and testing are the best ways to minimize these problems, but even with these preparations these sources of breakdown, malfunction, and backfire cannot be completely eliminated. New, untried technologies that are complicated in their field operation, or the use of unskilled or inadequately prepared soldiers or civilians to manage these technologies, may be especially conducive to this problem.

Legitimate Peripheral Interests of Initiators

The extent to which vital interests are at stake plays an absolutely critical role in analyzing casualty aversion strategies, with legitimate but remote security interests reducing the willingness to see body bags come home. The underlying assumption is that Western domestic populations are unable to tolerate casualties "when perceived core national interests are not at stake."[1] Of course, the notion of what constitutes vital interests is quite subjective, varying from society to society and across time and different predicaments; nonetheless, the perception of less-than-vital interests, including in military operations other than war, means that tolerance for human casualties would be at a minimum. Thus casualty aversion seems to work best when undertaking foreign military action involving less-than-vital interests: given the public's willingness to suffer casualties when vital interests are truly at stake, where "anticipating high collateral damage merely means that the military value of the target must be great enough to justify the unintended losses,"[2] the political payoffs would be greatest when such willingness to suffer is absent.

Although the interests precipitating the application of technologies useful for casualty aversion may be less than vital, they need to be legitimate for such application to be effective. In other words, the rationale for war or intervention involving these technologies ought to be quite explicit and capable of demonstrating to both domestic and foreign audiences that sound motives—preferably consistent with a preexisting doctrine or set of principles—are present for entering a military confrontation. Being able to demonstrate the legitimacy of one's action can indirectly help with minimizing loss of life, because this makes it less likely that a target population will resist or that an onlooker state will interfere.

Despite the often peripheral national interests addressed, casualty aversion strategies appear to be more effective when used by those who possess carefully formulated and articulated security objectives rather than by those who intentionally or unintentionally allow these strategies to serve as a pretext for action that is otherwise unwarranted. These well-specified military objectives also help to forestall the danger of foreign policy recklessness. In other words, war goals—including the presence or absence of clear criteria for success—"are likely to have a great deal to do with the willingness of the society to pay the human costs of war."[3]

Vulnerable Character of Targets

The psychology of a target also affects the utility of casualty aversion strategies. In particular, special difficulties associate with nonstate foes such as terrorists or insurgents. Adversaries who have passionate resolve for their causes or disregard for the value of their own lives, as opposed to conscripts who fight for pay or duty, pose difficulties. In many of today's intense religious and ideological struggles, targets seem to possess an advantage in zeal for their cause (asymmetric will) and appear willing to keep fighting and to endure unspeakable suffering against overwhelming odds. Highly irrational or unpredictable targets generally make matters worse, particularly if foes appear unable or unwilling to properly extrapolate current damage from precision-guided munitions, nonlethal weaponry, or information warfare to more widespread and devastating damage in the future; for reasons related to both cognitive and emotional limitations, adversaries may overestimate their own resilience in the face of significant losses, or alternatively underestimate the initiator's will to continue the attack, making outright obliteration appear to be the only way to settle a conflict.

On the other hand, a rational state without resolve for its cause would appear an ideal target for effective casualty aversion. This would appear particularly likely if a designated target interprets the initiator's restraint in minimizing loss of life as increasing the chances that the initiator's objectives are reasonable or legitimate. In other words, this demonstration of voluntary force limitation by initiators could signal to targets that the thrust is worth careful consideration.

Target vulnerability is a key casualty aversion consideration. Casualty aversion strategies appear to make more sense when enemies possess key identifiable vulnerability points exploitable through precision-guided munitions, nonlethal weaponry, or information warfare. Often, having vital targets and weapons systems highly centralized—largely located in a few locations—is quite helpful. The underlying logic is that rapid enemy capitulation can result from knocking out a few nodes critical to the economy or the armaments industry, with the surgical removal of an enemy's most vital elements crippling its capacity to fight and promoting faster surrender. Without target possession of such vulnerability points, innocent bystanders are more likely to be accidentally killed or injured, and the lack of massive brute force present within many technologies useful for casualty aversion may prove to be a critical handicap in achieving military victory.

Furthermore, an enemy adept at dispersion, concealment, and camouflage, able to move vital military assets quickly and without detection, reduces the effectiveness of casualty aversion strategies. Situations where adversaries are capable of eluding ready target acquisition—conventionally established through moving target indicators, penetrating radars, midcourse corrections, and prompt and responsive delivery systems—seem more likely to promote unintended noncombatant casualties. Generally, the greatest difficulties may emerge when dealing with targets possessing effective countermeasures that negate the advantages of technologies useful for casualty aversion in warfare.

Discrete and Restricted Nature of Missions

Casualty aversion strategies are clearly not appropriate for all types of foreign military missions. Casualty minimization values would seem to fit better with short-term missions with limited objectives than with long-term missions with grandiose objectives:

> The "no-casualties" mantra can work for purely punitive missions, where the job at hand consists of discharging cruise missiles from afar, just as in earlier days a gunboat could shell a native village at no risk to itself. But this no-risk mind-set can be fatal for any mission more ambitious. Somalia presents the worst-case scenario: Americans arrive full of vim and vigor, only to leave, tails between their legs, after 18 soldiers die. This sends a message of American irresoluteness that can only cause trouble later.[4]

Often, adherence to fixed time involvement, bounded aspirations, or restrained coercive means can be beneficial in this regard. Predicaments conducive to the quest for bloodless war thus might include, for example, inducing a hostile regime to cease its external aggression or decapitating enemy leadership. However, many military scenarios exist wherein such an approach would be totally inappropriate (including, for example, the United States unilaterally attempting to implant civil society in a foreign country): "Obviously, our new way of war is of limited value in situations requiring the conquest, occupation, and administration of territory; these missions require 'boots on the ground' in sizable numbers."[5] In a parallel way, minimizing loss of life seems more critical when used in situations to manage acute short-term threats rather than chronic long-term sources of instability.

Of particular interest is the general suitability of casualty aversion strategies in foreign military actions where humanitarian issues play a prominent role. The advent of an era of foreign military intervention for humanitarian purposes "has posed vexing new problems for government leaders," as such missions "do not allow for the use of overwhelming force to attain quick, decisive victories":[6]

> The starting point rules of engagement for such operations—as it is for counter-insurgency operations—is the imperative of utmost restraint and discrimination in applying force. Firepower is an instrument of last rather than first resort. There is no big enemy to close with and destroy, but rather the presence of threatened civilian populations that must be protected in a way that minimizes collateral damage.[7]

Prioritizing human protection thus appears to be most essential in volatile situations where casualties, particularly of noncombatants, are both politically unacceptable and militarily unnecessary.

The quest for bloodless war also works better when the mission objective in a military confrontation is not the complete eradication of the enemy but rather the achievement of influence over one's target. In such circumstances, it is important not to alienate unnecessarily the citizenry of a country one expects to hold sway over in the future. After conflicts end, casualty aversion strategies could make restoration of the attacked system faster, easier, and cheaper. Thus casualty aversion makes the most sense when there is an expectation that the majority of citizens of a target country can be "rehabilitated" after a change in government.

Capacity to Discriminate During Combat

Perhaps the most fundamental prerequisite for the successful pursuit of bloodless war is heightened intelligence requirements. Minimization of civilian casualties during wartime requires exceptionally accurate information about the location of enemy targets. Pinpoint targeting is crucial, and inaccurate intelligence can cause even the most casualty-sensitive state to find itself continuously killing innocent victims rather than hitting primary targets. Although electronic intelligence is important, the availability of human intelligence may be most pivotal in minimizing casualties because remote surveillance systems cannot accurately predict how enemies will react or easily distinguish (even with heat sensors) between a building packed with people and one packed with operating military machinery that constitutes a primary target. Caring about human casualties would clearly necessitate a much higher level of intelligence than would be the case otherwise.

Moreover, for casualty aversion strategies to have desired long-run impacts, an initiator needs to possess a full understanding of the culture and of the target groups' preexisting concerns and perceptions. Rather than being a classic intelligence acquisition issue, this need would entail education, training, and possibly even direct contact and experience with members of a target country. Often the identification of key vulnerability points is a function less of scanning for key military installations and more of understanding how authority in a particular society works.

In a related manner, the ability to distinguish between civilians and combatants can be a pivotal influence on whether casualty aversion is warranted. The distinction between combatants and noncombatants "is widespread in world history," and in Western culture "the ideas of noncombatancy and of noncombatant protection run deep in moral tradition,"[8] with the distinction between the soldier and the civilian traditionally based on who does and does not bear the moral guilt for waging and participating in the war.[9] Given the improved ability of modern weaponry to discriminate among targets, this differentiation deserves consideration in deciding whether to focus on casualty minimization precisely because of its difficult implementation in modern warfare. If existing political and physical restraints allow targeted groups easily to mix civilians and combatants together in a manner that confuses initiators, even the most zealous advocates of most casualty aversion strategies may find their application inappropriate. It is important to be able not only to identify civilians and minimize their loss of life, with a sharp and clear distinction between friend and foe and between targets and bystanders, but also to protect or avoid destruction of essential civilian services such as public health and education. Even though casualty aversion strategies are frequently well matched in other ways to third world intervention scenarios, the difficulty of isolating who is a combatant in many developing countries is a real obstacle to be overcome. Nonstate groups such as terrorists again pose particular problems here; in addition to encountering a mix of civilians and soldiers, when confronting such adversaries there is often no reliable way of identifying who is a combatant and who is a noncombatant, nullifying the general target differentiation capabilities of advanced weapons technologies.

Superiority of Advanced Casualty Aversion Technology

The quest for bloodless war is most effective when undertaken by those who possess superior overall military capabilities. Facing ruthless and aggressive expansion by a powerful enemy force, such as the United States experienced with regard to both Germany and Japan during World War II, would not generally seem to pose the best circumstances for being sheepish about mowing down foes. In such circumstances, the aforementioned dangers of being perceived as weak, or even appeasement-oriented, would certainly surface. Similarly, if aggregate power ratios were such that one's country is at a severe disadvantage and lacks credibility, the timidity sometimes signaled by casualty aversion could be debilitating. In contrast, having an overall power advantage over one's foe can allow the application of technologies useful for casualty aversion to serve as a somewhat subtle reminder to targets of the vastly greater force that could be applied if they continue to resist.

Casualty aversion seems especially effective when initiators possess an asymmetrical advantage in their ability to utilize precision-guided munitions, nonlethal weaponry, and information warfare, as well as effective countermeasures. Otherwise, an initiator could easily face a surprising and potentially devastating level of intrusive retaliation, leading to a two-sided high-tech confrontation where a decisive outcome would be elusive. Because of the nature of the technologies, such an action-reaction cycle could easily leave both societies' infrastructures devastated. Ideally there would not be other states interfering with the dominance of the casualty-minimizing initiator, either by trying to influence the target in contrary directions or by using casualty-maximizing techniques to achieve influence in the area.

Technologies useful for casualty aversion appear to work best when there is no ambiguity about either their presence or an initiator's will to use them in a particular conflict. Doubt about such capabilities can only cause potential targets to test whether they exist by undertaking disruptive action. In contrast, explicit knowledge of these technologies by an adversary has the potential to make their use less urgent or even ultimately unnecessary.

As war becomes more technology-intensive, with everyone realizing just how great a role advanced weapons systems play, killing of humans—combatants or civilians—may accomplish less and less. In such conflict, managing to kill 100 or even 1,000 enemy soldiers may do little to achieve ultimate victory, especially against a casualty-insensitive target, compared to demolishing a military structure or command-and-control database. In such situations, casualty aversion strategies may emerge as a practical necessity, with the maintenance of superiority in relevant technologies ever more crucial.

Isolated and Integrated Thrust of Action

The containability of a coercive confrontation is an important determinant of the utility of casualty minimization. Often, modern wartime combat cannot be easily delimited, either within states or across states, causing major difficulties in imposing restrictions related to the loss of human life. Protecting civilians, for example, is easier to undertake in readily demarcated, predetermined areas. Thus casualty aversion appears to have the greatest utility when undertaken in more localized warfare whose bounds remain fixed than in conflict that exhibits contagion and cannot easily be confined.

The effectiveness of casualty minimization strategies depends on their seamless coordination with other foreign security policies in a particular confrontation. Casualty aversion appears to work best when combined with other traditional military force strategies, integrated fully with other military and political initiatives to disrupt the target, than when used alone. An unfortunate and clearly suboptimal past pattern has been to have technologies like precision-guided munitions, nonlethal weaponry, and information warfare become quite disconnected in their application from conventional military strategies.

Ideally, minimizing loss of life should also involve extensive coordination among military units and services and civilian information agencies. Because there is a tendency for a particular service to focus on a particular technology useful for casualty aversion—such as the U.S. Air Force with precision-guided munitions or the U.S. Marines with nonlethal weaponry—gaps have existed in this coordination and the prerequisite mutual understanding. There is still considerable resistance in the United States and elsewhere for different services and units within the armed forces to abandon their separate cultures and traditions to undertake joint operations involving incredible coordination without one group clearly dominant or in charge. The goal here is to combine coercive strategies in such a way that will not only minimize the casualties inflicted by an initiator but also restrain a target state's aggression because it fears retaliation from an initiator during the war or after it is over. Thus the combination of sound integration of casualty aversion strategies into overall foreign security policy with coordinated action across military services can be a crucial signaling device to adversaries of strength and resolve.

Juxtaposing Conditional Utility to Global Threats

Having outlined in a preliminary way the circumstances where the use of casualty aversion strategies seems the most prudent, it seems important to compare the conditions identified for optimal application to those that actually pervasively characterize today's global threats. One could easily imagine a dysfunctional scenario where the means of undertaking the quest for bloodless war are highly effective in a wide variety of circumstances but unfortunately not in those where the application of force is most needed.

Regarding the well-defined motivations of initiators, the match is mostly though not completely favorable. In the case of the United States, for instance, many of the international dangers conducive to foreign military action represent not direct threats to its continued existence or national security but rather indirect and secondary threats to international stability, and this situation matches nicely with the best case for casualty aversion. However, the precise specification and articulation of rationales for military confrontations is often inadequate.

Regarding the vulnerable character of targets, the pattern is bleaker. Most of the targets that the United States and other major powers perceive today as the greatest sources of threat—violent religious fundamentalists, ruthless transnational terrorists, and probloodshed extremist groups of all stripes—are highly passionate for their causes, willing to die fighting for them, and not prone to perform the kinds of cost-benefit analysis that the West uses to define rationality. The utility of casualty aversion strategies is decidedly low under these circumstances; as one analyst asks, "How, therefore, can armed forces, staffed by professional, salaried, pensioned, and career-minded military personnel who belong to a nation intolerant of casualties, cope with aggressors inflamed by nationalism or religious fanaticism?"[10] As to the existence of pivotal fixed identifiable targets, these generally exist in the rogue states, making casualty aversion workable there, but are often absent or very hard to find in unruly nonstate transnational groups.

Regarding the discrete and restricted nature of missions, findings are again disappointing. For example, the United States has recently undertaken military actions that at first appear to be short-term with limited objectives but then later turn out to be quite different; the attempts to develop stable democracy in Afghanistan and Iraq come to mind immediately. Furthermore, as to achieving influence over a target that can be rehabilitated, the United States believes this to be the case for most rogue states—even for Iraq, Iran, and North Korea the perception is that the governments have been corrupt or evil but the citizenry can be redeemed—but decidedly not for terrorist groups.

Regarding the capacity to discriminate during combat, an even greater mismatch exists between the strengths of casualty aversion and the nature of existing threats. While the West's electronic intelligence is generally quite advanced, its human intelligence about the most important sources of international instability is relatively weak, given the elusive nature of the threats. As to the ability to distinguish between civilians and combatants, virtually every one of the enemies of the United States has learned to place vital military targets near schools and hospitals, to disguise combatants and those engaged in suicide strikes so that they look exactly like innocent civilians, and to disperse soldiers in heavily populated areas so that any attack will generate significant collateral damage. Even though this cluster of conditions can characterize urban combat, for which some nonlethal weapons and psychological operations were originally designed, it does not constitute an optimal scenario for the general application of casualty aversion strategies.

Regarding the superiority of advanced technologies useful for casualty aversion, here is the one area where there is a near-perfect match between the advantages of the quest for bloodless war and the nature of existing threats. Western powers maintain a huge superiority in overall military capabilities over their enemies, consisting mainly of terrorist groups and rogue states. Moreover, the United States in particular possesses a huge asymmetrical advantage in its highly sophisticated and reliable precision-guided munitions, nonlethal weaponry, and instruments of information warfare over its adversaries, with unintended diffusion of these technologies to unruly parties currently at a bare minimum. Similarly, thanks both to very clear pronouncements and warnings from the U.S. government and to actual successful application of these technologies since the 1990s, there is absolutely no doubt in the minds of potential adversaries that the United States has effective technologies useful for casualty aversion and would use them in combat without any hesitation.

Finally, regarding the issue of isolated and integrated thrust of action, here again the pattern is less than optimal. Many of the threats facing the West today are not neatly localized and containable but instead are contagious, as exemplified by the slight spillover into Pakistan of the 2001–2002 terrorist threat in Afghanistan, and the slight spillover into Syria of the 2003 threat in Iraq. Anytime the United States, for example, confronts a primary enemy, there are sympathizers in neighboring countries who are willing to provide safe haven for those attacked and who are in many cases willing to join the resistance themselves. Thus the rules of engagement may keep changing, as what was formerly a bystander or a neutral party suddenly becomes a new threat (in the 1990s, this was evident at various times among the different ethnic factions residing in the former Yugoslavia). However, on a brighter note, there has been considerable progress in integrating casualty aversion with other military strategies and in coordinating the participation of different services and units in casualty-minimizing efforts.

Conclusion

Unfortunately, this brief analysis demonstrates a significant (though not uniform) mismatch today between the security needs posed by international threats and the security advantages provided by casualty aversion strategies. With the exception of one category—advantages in advanced casualty technology—the threats often appear to expose the weak underbelly of the quest for bloodless war. In particular, casualty aversion strategies generally do not seem well suited to confront non-state transnational threats such as terrorist groups that appear not to be susceptible to rehabilitation, not open to Western-style rationality and cost-benefit analysis, not in possession of vulnerability nodes, not vulnerable to penetration by conventional forms of intelligence, not easily distinguishable from innocent civilian populations, and not isolated from support by sympathizers and even formally organized "cells" outside of the immediate target area. In the short run there seems to be very little that Western states can do to dramatically alter these shortcomings.

What this means is decidedly not that states ought to discard the quest for bloodless war because its strengths do not always fit well with today's most ominous foreign predicaments, but rather that security policymakers ought to view casualty aversion strategies as just one of many useful sets of instruments to pursue national interests abroad. In the future, of course, it is likely that the array of international threats will change, and thus the value of casualty aversion strategies could change relative to other approaches. So many of the dangers the West faces around the world today require a combination of being resolute and flexible, as well as being seen as both resolute and flexible; no state desiring to project these qualities would want to be seen as following a lockstep formulaic approach to war where the real political-military objectives become obscured by a monomaniacal obsession with force protection or civilian casualty minimization. What appears to generate admiration and credibility in the international arena may thus not be so much the ability to develop and apply advanced technologies useful for casualty aversion but rather the ability to discriminate prudently between situations where their application as well as the use of bloodier coercive strategies is most and least warranted.

Notes

1. Piers Robinson, *The CNN Effect: The Myth of News, Foreign Policy, and Intervention* (New York: Routledge, 2002), p. 40.

2. Colonel Charles J. Dunlap Jr., "Law and Military Interventions: Preserving Humanitarian Values in Twenty-First-Century Conflicts" (Washington, DC: Harvard University, John F. Kennedy School of Government, Carr Center for Human Rights Policy, Humanitarian Challenges in Military Intervention Conference, November 29, 2001), p. 9.

3. Scott Sigmund Gartner and Gary M. Segura, "War, Casualties, and Public Opinion," *Journal of Conflict Resolution* 42 (June 1998): 298.

4. Max Boot, *The Savage Wars of Peace: Small Wars and the Rise of American Power* (New York: Basic Books, 2002): p. 347.

5. Jeffrey Record, "Collapsed Countries, Casualty Dread, and the New American Way of War," *Parameters* (summer 2002): 20.

6. Karl W. Eikenberry, "Take No Casualties," *Parameters* (summer 1996): 112.

7. Jeffrey Record, *Failed States and Casualty Phobia: Implications for Force Structure and Technology Choices*, Occasional Paper no. 18 (Montgomery, AL: Center for Strategy and Technology, Air War College, September 2000), pp. 19–20.

8. James Turner Johnson, *Morality and Contemporary Warfare* (New Haven: Yale University Press, 1999), p. 125.

9. Pauline M. Kaurin, "Innocence Lost: The Future of the Combatant/Noncombatant Distinction," Joint Services Conference on Professional Ethics, 2002

(unpublished paper), available online at www.usafa.af.mil/jscope/jscope02/kaurin02.html.

10. Edward N. Luttwak, "Toward Post-Heroic Warfare," *Foreign Affairs* 74 (May–June 1995): 115–116.

8
Security Policy Implications

The lessons from the preceding conceptual analysis, combined with recent wartime experiences, point to some preliminary directions for foreign security policy in light of the quest for bloodless war. Taking into account the specific circumstances associated with the utility of casualty aversion, more general security prescriptions, summarized in Figure 8.1, emerge from this analysis. Given a continued concern with minimizing loss of life, how can we channel this noble impulse in more functional ways than in the past?

Obtain Better Background Information

To begin with, there needs to be more rigorous conceptual formulation and empirical testing of the opportunities and restraints posed by minimizing casualties during warfare. Although this book provides an important step in this direction, carefully delineating the areas of concern and laying out deductively the probable areas of greatest danger and greatest utility, it obviously cannot definitively answer all of the major questions surrounding the complicated quest for bloodless war. Future research could isolate more exactly when the strategies associated with this quest are most and least useful as part of a war-fighting package. Resource allocation in the security policy setting based on this kind of analysis ought to focus at least initially on those facets of a predicament where transformation is most possible rather than on those facets historically resistant to change (such as the basic value orientations of target societies).

Figure 8.1 Security Policy Advice and the Quest for Bloodless War

Obtain Better Background Information
Undertake more systematic research on the conditional utility of casualty aversion.
Gather better international intelligence, particularly human intelligence on terrorists.
Reevaluate the delicate balance between military and political priorities.

Recognize Limits of Casualty Aversion Technologies
Understand the clear limits of casualty minimization strategies.
Accept that a continued casualty aversion focus restricts global hegemony.
Resist the temptation to be reckless in the use of force.

Consider Military/Civilian and Domestic/Global Norms
Be sensitive to prevalent values on civilian and combatant casualties.
Consider the changing nature of the global security environment.
Restore and reinforce the national military ethos.

Better Communicate Goals and Expectations
Explain more clearly the rationale for war or foreign military interventions.
Clarify expectations and achieve more stable long-term conflict outcomes.
Acknowledge how casualty aversion affects the image of strength and resolve.

Modernize, Diversify, and Protect Battlefield Strategies
Improve casualty aversion technologies and cross-military service implementation.
Diversify battlefield strategies including unorthodox, "outside-the-box" solutions.
Guard against unintended diffusion of technology and develop countermeasures.

Some important variables that follow-up studies might consider are the impact of casualty aversion strategies on the frequency and duration of war (and on the ease of reaching settlements that end violent conflicts); the way the power disparity between opposing sides (and the involvement or noninvolvement of the United States) affects an emphasis on casualty aversion; the reliability and predictability of long-term and short-term application of casualty aversion strategies; the effectiveness of enemy countermeasures and ability to exploit one's focus on casualty minimization; the relative value of technologies useful for casualty aversion versus that of indiscriminate brute-force technologies; the way the quest for bloodless war interacts with the probability of successful deterrence; and the military and political success or failure of coercive overseas missions emphasizing casualty aversion. Such future efforts should help to nail down more precisely the relative value of utilizing casualty aversion means in the conduct of warfare.

Furthermore, throughout this study it has been consistently evident that one of the most essential areas for minimization of casualties during wartime is improved foreign intelligence, particularly human intelligence on nonstate targets like terrorists. Among the types of intelligence needed are better information about the location of enemy targets, including improved ability to isolate and unambiguously identify primary vulnerability points, and better understanding of casualty aversion strategies' short-term and long-term impact on particular types of predicaments. Policymakers need to consider unusual possibilities in this regard, including the use of unorthodox sources such as transnational criminal organizations, to get the kinds of information on terrorists and violent religious fanatics prerequisite to the pursuit of bloodless war.[1] Yet this intelligence challenge keeps growing increasingly complex: "There are just too many haystacks in which to hide; our reconnaissance drones will surely get much better, but so will the enemy's disguises."[2]

Recognize the Limits of Casualty Aversion Technologies

Armed with this better background information, a fuller delineation of the limits of casualty aversion strategies is needed. No magic silver bullet exists to facilitate the attainment of the quest for bloodless war:

> No obvious cure exists for the affliction of casualty phobia. Hopefully, the elites themselves will come to recognize that the public's tolerance for casualties is much more contingent than pessimists believe or want to believe, and that the political leadership can greatly influence public attitudes on casualties in a given situation. Given the strength of the elite's conviction that the people they serve have little stomach for war under almost any circumstances, however, it would probably take an actual demonstration of casualty tolerance to change minds. But this hardly means seeking another war just to prove a point. Moreover, the United States is fast running out of enemies capable of inflicting significant casualties on deployed US military forces.[3]

Thus, in advance of warfare, we need clarification of exactly what we do and do not expect casualty aversion strategies to accomplish.

We also require a more refined comprehension of casualty aversion strategies' range of application and their costs and benefits, including the perception of these strategies and their effectiveness and legitimacy by domestic and foreign civilians and military personnel as well as by government officials. To help inform these views, both the domestic public and foreign onlookers need to be educated about these limits and dangers and about the moral and practical trade-offs associated with short-term casualty minimization. The goal should be for all concerned to develop an understanding that precision-guided munitions, nonlethal weaponry, and information warfare are tools of selective utility, not cures for every international threat that is likely to emerge.

Given the limits of casualty aversion strategies, the United States in particular should take a fresh look at its international military obligations:

> If Americans cannot adopt a . . . bloody-minded attitude, they have no business undertaking imperial policing. I am not suggesting that Americans prepare themselves to suffer thousands of casualties for the sake of ephemeral goals, but policymakers should recognize that all military undertakings involve risk and should not run away at the first casualty. More important, Washington should not structure these operations with the prime goal of producing no casualties. This is a recipe for ineffectiveness. . . . If the U.S. is not prepared to get its hands dirty, then it should stay home.[4]

Commitment usually necessitates a willingness to consider undertaking unrestrained use of force against the adversary. Either the United States should be capable of persuading the public that military objectives in foreign wars or interventions are worth fighting for and risking lives, or "perhaps we shouldn't be undertaking these missions."[5] In many ways "we may call for a crusade to expand tolerance and democracy in the world, but do not have the stomach for the slaughter that such a crusade requires."[6] It is certainly hard to play the role of world policeman while being reluctant to carry a big stick: "Aspiring to be both global hegemon and righteous democracy, the U.S. has struggled with the dilemma of using its vast power while satisfying the self-imposed requirement that it act in a morally defensive manner."[7]

The United States needs to avoid any tendency—based on the questionable assumption that few American lives will be lost—to initiate too many wars, intervene militarily in too many areas of the world, or allow ongoing conflicts to persist for too long. One critical safeguard may involve a different kind of cost-benefit analysis, one that goes well beyond casualty minimization to look at other kinds of losses that may occur as a result of being involved in too many disparate battles. The temptation for the United States to go someplace where it has minimal national interests to set things straight and leave quickly without bloodshed, such as was recently contemplated regarding Liberia, just does not appear to work well in today's complex global security environment.

Consider Military/Civilian and Domestic/Global Norms

In any country's reconsideration of war-fighting strategies, international security norms pertaining to civilian and combatant loss of life merit attention. Currently, states and transnational groups vary widely in their beliefs about casualty minimization during war and exhibit little movement toward homogenized security values, and so nothing close to an overarching consensus exists. Some deeply divisive value differences, such as those revolving around the choice between stability and justice and between authority and anarchy, serve to impede sound policy decisions about whether presumably worthy goals justify substantial loss of life during warfare. The level of optimism exhibited by observers about whether progress can occur in reducing these divisions rests to some extent on their assumptions about the nature of the global security environment. Some assert that we have returned to a condition of impending foreign peril parallel to the early days of the Cold War, while others contend that the post–September 11 fears represent a temporary insecurity blip. More frank discussion should occur within and across societies about these differences in relation to casualty sensitivity, incorporating a combination of top-down and bottom-up ideas, with the aspiration of at least creating greater understanding of the diverse perspectives if not fostering shared values.

Given this diversity of viewpoints, it is inappropriate today for the United States, any other great power, or the United Nations to act as if consensus did exist or—even worse—to impose its interpretation and rules pertaining to casualty aversion on others, even among states claiming to be democracies. The presumption that most of the world has similar beliefs about the acceptability of human suffering during war seems downright naïve. Moreover, attempting to mandate common values on casualty aversion through pressured international agreements is clearly not the answer: "Law is not—and can never be—the vehicle to ameliorate the horror of war to the extent its advocates hope and, indeed, seem to expect."[8] However, should the carnage associated with international anarchy at some future point lead toward a naturally evolving harmony on casualty aversion norms, countries should be attentive and meaningfully adapt their mix of offensive and defensive strategies during wartime through transparency, accountability, and possibly even codes of conduct.

Making security decisions to employ casualty aversion strategies is intertwined with not only external global but also internal domestic moral standards. Such action raises the lofty question of whether the quest for bloodless war and its associated means are compatible with democratic—and U.S.—purposes and principles. Different societies may, of course, answer this question in different ways. For some, the preservation of democracy and maintaining internal freedoms entail placing such a high value on the lives of citizens that protecting them on and off the battlefield virtually always becomes the top priority, while for others, the preservation of democracy against external threats is so important that virtually any sacrifice, including massive loss of life, is worth the cause. Given domestic differences in perspective among government officials, military officers, and the mass public, there should be greater expectation and acceptance of sustained internal disagreement—and even heated friction—about crucial casualty aversion issues surrounding the proper use of force during warfare.[9]

U.S. armed forces themselves have some work to do to restore and reinforce (and possibly positively transform) the military ethos in the United States in such a way that the timidity often associated with prioritized force protection does not become the dominant image either of the nation as a whole or of the soldier in particular. Service academies and training programs need to reemphasize courage and willingness to sacrifice. The addiction to new, powerful technologies that keep soldiers comfortably out of harm's way needs tempering from values like "duty, honor, and country" that entail the willingness to place one's life in direct peril so as to achieve victory in warfare. The military message of "whatever it takes" needs to stay separate from the political message of "costless" war.

Communicate Better Goals and Expectations

In the face of both internally and externally diverse and conflicting security perspectives, those initiating the use of force in a foreign military confrontation need to explain more clearly the rationale for engagement both to their citizenry and to foreigners. There is a well-established pattern that, as uncertainty about mission objectives rises, tolerance for casualties falls, and this means that communicating convincingly the motives and goals for a foreign intervention or war is absolutely necessary in order to preserve choice and flexibility in military action. In addition, it is important to articulate preexisting limitations affecting action, including willingness to sacrifice human life both during a violent conflict and in the aftermath of such conflict. Moreover, in the anarchic post–Cold War setting, a convincing articulation of mission is important because of the inherent difficulty of establishing a clear basis for legitimate coercive action overseas. More specifically, initiating governments need to carefully specify political and military objectives and national interests in each case, avoiding a reliance on justifying action through casualty minimization alone, in the process increasing prospects for internal and external support for their security initiatives. This explanation should include not only short-term goals but also long-term expectations about stable conflict outcomes. The simple concept that military victory virtually and inevitably requires sacrifice of human life needs explicit articulation and defense.

Through this careful exposition of rationale, the U.S. government in particular needs to communicate to its allies and citizens its sense of prudent restraint, and to its adversaries its firm sense of resolve in military confrontations. Taking advantage of the emotional rage of many U.S. citizens due to the terrorist attacks of September 11, 2001, should not form the cornerstone of any such justification for action. Part of this communication should incorporate an explanation that all the primary means of pursuing bloodless war—including precision-guided munitions, nonlethal weaponry, and information warfare—can indirectly generate human casualties; otherwise, overblown expectations about casualty minimization can result in horror and outrage when even a few people die in combat. The message should include not only that warfare itself is often both bloody and prolonged but also that the aftermath of such warfare is often rocky, as exemplified by the extensive killing of U.S. soldiers that occurred after the end of the 2003 war against Iraq.

A seemingly inescapable tension emerges in this regard for U.S. security policy:

> The larger political issues involved in attempting to find a balance between casualties and policy objectives are . . . intractable . . . and will always be particularly acute in liberal democracies. However, it should be pointed out that one of the most difficult dilemmas facing our statesmen today is how to respond effectively to domestic concerns about losses in conflicts abroad, while still showing the tangible signs of commitment necessary to maintain a claim to coalition and world leadership. With increasing US participation in multinational military operations likely in the years ahead, the subject will probably become more contentious.[10]

On the one hand the United States needs to convey to its foes that it is willing to endure great losses and continue the battle as long as it takes to emerge victorious, while on the other hand the United States needs to convey to its own people that it will do everything possible to minimize loss of life and get out of any foreign coercive entanglement as quickly as possible.

As part of this process, the United States needs to consider more carefully how its use of casualty aversion strategies affects its internal and external image. How concern about loss of life or injury affects the overall image of a state's strength and stability in the eyes of onlookers, particularly those posing a potential threat, has a dramatic impact on long-term security. Internally, security policymakers need to avoid projecting an image of manipulation of casualty aversion simply to mollify the people or to camouflage other, less noble purposes. Externally, in light of the common concern that body bags coming home reinforces an image of weakness, some method needs to be developed to disconnect or at least distance a state's willingness or unwillingness to sacrifice human life in the pursuit of foreign security policy from its international credibility. The continued presence of such a tight linkage potentially creates a "lose-lose" proposition for the United States. If it chooses to respond to this unfavorable image and engage in bloody battles all over the globe, then it risks becoming overextended and achieving long-term goals in none of them; yet if it chooses to ignore this unfavorable image and continue to minimize loss of life, then it risks losing respect among allies and fear among adversaries.

Modernize, Diversify, and Protect Battlefield Strategies

There needs to be increased diversification in battlefield strategy and tactics and in funding of weapons systems to prepare for a variety of both bloody and not-so-bloody confrontations, with a recognition that one approach does not fit all coercive scenarios. This diversification includes openness to new approaches to warfare that may prove to be functional in the future, such as utilizing heavily automated military units or employing private foreign soldiers to fight in battle; the key appears to be to weigh dispassionately the costs and benefits of each new possibility, and then to integrate carefully the ones that pass muster into overall military strategy. There needs to be a careful rethinking of the balance between casualty-minimizing strategies and casualty-maximizing strategies, and between achievement of victory and protection of life, from both military and political vantage points. The rules of engagement themselves may merit conceptual reconsideration and increased flexibility in application. The underlying guiding principle here ought to be versatility of means to cope with different kinds of challenges rather than exclusive or even heavy reliance on technologies useful for casualty aversion to manage all types of military confrontation.

Given the increased emphasis on minimizing loss of life during warfare, it might be worthwhile to think about how strategy and tactics in recent military confrontations—and their associated outcomes—might have differed had not civilian casualty aversion and force protection been so high a priority on the political agenda. From this kind of analysis, perhaps one could determine what kinds of underutilized and underfunded approaches merit more sustained consideration for possible future use in tandem with those associated with the quest for bloodless war. Although support ought not to be withdrawn from traditional means of casualty aversion, in the future we need to develop a robust and carefully tuned mix of well-tested approaches to use in the battlefield.

For the toughest threats not readily subject to change, such as dealing with passionate terrorists, ruthless wily dictators, or extreme, violent religious fundamentalists, considerable creativity and "out-of-the-box" thinking are essential, for even the most diligent application of standard military operating procedures has not adequately accomplished the job. Because volatile parts of the third world are experiencing the opposite of the growing casualty sensitivity among advanced industrialized countries, we need different means of casualty aversion toward these targets, different ways to apply existing means, and—most likely—recognition that minimizing loss of life should not invariably be a top priority against this type of target.

For the United States, one embedded paradox is that, if it continues to be heavily guided by casualty aversion, it may be diverted from ever discovering the benefits of alternative military strategies: "Domestically, continued acceptance of the quick, costless war presumption risks becoming a debilitating, self-fulfilling prophecy; after all, how can we expect the American public and military to cope with future exigencies requiring protracted operations or entailing significant casualties if such contingencies are defined away as beyond American will?"[11] Some military and political leaders do not want to risk the consequences of opening up the Pandora's box of the full range of war-fighting options, and their somewhat misplaced caution may prevent their countries from ever seeing the long-term lifesaving impact of short-term, casualty-intensive military strategies.

Even if the United States were to escape from overreliance on casualty aversion, it would still need to resolve the delicate tension between military and political priorities:

> In making decisions to use—or not use—military force, presidents must weigh considerations of both military effectiveness and domestic politics. The United States is, after all, a democracy, and, contrary to conventional wisdom, politics never stopped at the water's edge. Yet if perceived domestic political considerations become the enemy of military effectiveness, to the point of arbitrarily excluding use of one form of military power in its entirety—and this is the direction where America unfortunately seems to be headed in the present era of small wars in out-of-the-way places—then the United States must alter established force structure and patterns of defense spending to maximize the effectiveness of the forms of military power it is prepared to use.[12]

Regardless of the resolution, the determination of this balance needs to be more dynamic—highly sensitive to changing security circumstances—for no simple dogmatic solution can serve to resolve these inescapably thorny issues.

However, as long as technologies useful for casualty aversion continue to be employed in modern warfare, those using them need to continue to improve and refine them to maximize their effectiveness and minimize their risks. This modernization should specifically involve efforts both to avoid accidentally hitting unintended targets and to prevent hits on intended military targets from accidentally causing collateral damage in nearby civilian areas. Guidance systems and "smart" artificial intelligence capabilities need refinement to improve the ability to discriminate between combatants and noncombatants and between crucial military targets and peripheral installations. Moreover, effective application of casualty aversion strategies such as precision-guided munitions, nonlethal weaponry, and information warfare necessitates possession of a clear advantage over any potential adversaries. Although current technologies useful for casualty aversion have been working reasonably well, there is still a lot of room for improvement.

Armed with a wide and versatile range of strategic instruments for use in military confrontations, initiators need to make sure that casualty aversion strategies are fully integrated with other strategies—and fully coordinated across the different military services—to maximize chances of effectiveness. This entails much more extensive efforts at conceptualizing and testing how casualty-minimizing and casualty-maximizing coercive strategies might work harmoniously together in future confrontations. Preparation for this integration and coordination could occur, for example, through training exercises or war games focusing on applying technologies useful for casualty aversion in foreign coercive efforts. Despite the complex technologies involved, seamless coordination seems both necessary and achievable.

Assuming the means of pursuing bloodless war continue to be popular instruments of warfare, much work remains to safeguard against possible seepage of advanced casualty-minimizing technologies into the hands of undesirables abroad to be used against the United States in the future. Right now the United States has such a commanding lead in this area that it justifiably does not appear to be very worried about this issue, but that may change in the future. Despite their inherent orientation toward saving lives, any technologies useful for casualty aversion can be misapplied in the wrong hands to cause immense amounts of pain and suffering among target populations, including those of the state that originally developed the technology. So saving lives overseas could in the long run indirectly lead to more vulnerability at home—there is no such thing as an intrinsically sanitary weapons system that cannot possibly incur undesired carnage no matter who is operating it.

As a key element of this concern about unintended diffusion of technologies useful for casualty aversion, Western states should devote more attention to developing effective countermeasures. The need here is quite urgent. For example, currently the United States is vulnerable to both large quantities of long-range precision-guided munitions and multipronged information warfare efforts initiated from abroad. The availability of such countermeasures, if developed and demonstrated properly, could ideally communicate to others the futility of trying to use the West's own advanced technologies against it.

Finally, deserving considerably more attention are the consequences of the quest for bloodless war on the aftermath of violent conflict, including both reparations and restoration of the defeated country. Since casualty aversion strategies reduce carnage, presumably the anger and resentment following a military campaign that resulted in minimal loss of life would be less than otherwise. But nonetheless, such strategies leave alive and well many of those people within a vanquished society who are sympathetic to the enemy's cause, making postconflict return to normalcy a rocky road. Facing this predicament, the victorious state needs to develop new political means of taking advantage of this benefit while minimizing this cost.

Conclusion

Two conceivable future scenarios are clearly undesirable with regard to the quest for bloodless war. The first is that hysteria about potential abuse of precision-guided munitions, nonlethal weaponry, and information warfare might grow to such a degree that public research and discussion on the subject largely cease and domestic and international use is largely banned. Such drastic action would seem to be equivalent to "throwing out the baby with the bath water," where a potentially security-enhancing and livesaving instrument is rejected just because it can be misused. In any case, such a rash move appears likely to lead to a black market in the development and use of precisely the technologies that a ban like this would target. In contrast, the second scenario is that enthusiasm about the potential utility of this type of armament might escalate and proliferate so rapidly that virtually every group and state develops or acquires and indiscriminately applies significant capabilities for precision-guided munitions, nonlethal weaponry, and information warfare (or newer means of casualty aversion such as implementing automated robotic warfare or using private foreign soldiers to fight wars). Such global saturation creates the specter not only of heightened international insecurity but also of total negation of the turmoil management potential of these unorthodox coercive instruments.

At its best, casualty aversion can accommodate the widespread moral and democratic imperatives to keep violent conflict within acceptable bounds. Helping in this effort is the practical recognition that killing the largest number of people possible does not generally win today's wars. Certainly there have been many periods in human history when warfare was relatively "civilized"—fought according to a strict set of rules, restricting casualties to soldiers on the battlefield, and ending the moment political objectives were achieved—without the presence of excess carnage against either combatants or civilians. A combination of the speed, precision, and incapacitation of casualty-minimizing technologies ought to cause those contemplating overturning the international status quo to think twice about initiating such disruptive action.

At its worst, the quest for bloodless war can dramatically skew decisions about both when to intervene abroad and what military strategies to apply in foreign confrontations. It can lead to crippled military might and endless imprudent conflicts. It can distort and obscure the true nature of military missions, reducing the likelihood of effectively accomplishing key security objectives. The U.S. government in particular needs to consider carefully the seeming incompatibility between the grandiose political ends it wishes to pursue and the limited military means it is willing to apply.

If there is no change, and the United States and other Western powers continue to increase their reliance on casualty aversion strategies—without adequate reflection and selective limitation—as the primary way to fight wars, the results could be dire indeed. Enemies would simply wait the United States out, and allies would not want to become its coalition partners. The public would not be pleased, despite seeing fewer body bags coming home, because they would not witness a clean, decisive victory wherein all objectives are unambiguously attained; in other words, little would be lost in terms of costs, but at the same time little would be gained in terms of benefits. The military would obediently carry out missions as ordered, but morale would be low because of recognition on the part of many officers that the United States is not fighting the best fight. A country employing sophisticated casualty aversion strategies could make a big splash whenever it enters a confrontation, and could appear to have won a military victory, but then later not achieve its political objectives because of enemies' knowledge that it has no staying power. Adversaries will not even begin to keep up with the West in the latest advancements in precision-guided munitions, nonlethal weaponry, and information warfare technologies, so it will maintain an advantage here. But in a sense these foes will not have to keep up, because their asymmetric willingness to sacrifice life would still give them what they at least perceive as the advantage in the long run. The net result could be that the West may squander its superior military capabilities, fighting with one hand behind its back and rarely achieving long-term success.

Ultimately, is the quest for bloodless war an attainable aspiration, or is it equivalent to Don Quixote fighting windmills? Just as there is "no such thing as a free lunch," many analysts believe that there is no such thing as a "costless" military victory. Maintaining freedom requires sacrifice, and someone has to pay the price. Back during the U.S. Civil

War, Ulysses S. Grant taught the United States a valuable lesson—"that war is about death, pure and simple,"[13] and this insight is as valuable as ever:

> In any combat operation, blood will be shed, not only on the U.S. side but also among the enemy. This is an elementary truth, yet it was often overlooked in the 1990s when cruise missile strikes were planned to hit buildings after their occupants had gone home. Despite advances in modern weaponry, war can never be a clean, surgical business. Especially not urban or counterinsurgency warfare of the sort the U.S. finds itself undertaking in places ranging from Somalia to the former Yugoslavia. Sometimes . . . it can be a very ugly business indeed.[14]

Neither attempting to shield one's soldiers in the battlefield from experiencing harm, nor trying to handcuff them to keep them from using particular kinds of force against the enemy, seems very feasible in modern international conflict. Similarly, attempting to alter dramatically through military force a target society or regime without inflicting some level of suffering on its soldiers or its citizens seems to be a questionable proposition, given that fear of future damage may be unfortunately central to both surrender and progressive transformation. Although over the centuries the ways of fighting war have changed dramatically, there is little doubt that its basic character has always been and will always be bloody.

The primary explanation of the current improbability of costless victory involves the kind of objectives pursued in modern military confrontations. Generally, the purpose of the fighting is not simply to win the battle or to dominate a piece of territory, but rather to change the way a group or country operates in the long term. Often, the ultimate goal is the creation of some form of democratic civil society in the area. This kind of objective almost automatically means that a quick surgical strike at an adversary—no matter how carefully tuned or technologically sophisticated—will not be sufficient to trigger desired long-term changes. The United States and the West have been very adept at using this speedy casualty-minimizing strategy to stop offensive behavior and remove offensive leadership, but not to install effectively a replacement set of attitudes and customs that create an open system with stability over the long haul.

The "no pain, no gain" mentality opposing casualty aversion in warfare certainly does not mean that it is never prudent to seek to find ways to reduce bloodshed—many important circumstances certainly exist under which this makes sense—but it does mean that the normal expectation of governments and their citizenry should be that significant loss of life will continue to accompany (despite the availability of technologies useful for casualty aversion) achievement of military objectives in coercive confrontations overseas. It is entirely possible for a country with vastly superior armaments and training to win a war with virtually no loss of life, but the ability to achieve long-term political objectives in such situations is severely in doubt. Perhaps the best testimony to this pattern is the difficulty the United States is having today is establishing a stable participatory democracy in Afghanistan and Iraq after swift and relatively casualty-free military victories in both countries.

In the end, a critical need exists for an informed, restrained, norm-sensitive, articulate, and diversified approach to modern warfare, in which resort to technologies useful for casualty aversion occurs only in a manner carefully integrated with other approaches in those carefully defined circumstances that lead to their most effective use in accomplishing overseas missions. Maintaining openness to a fluid mix of strategies for managing different kinds of threat, rather than being rigidly wedded to those strategies that minimize or maximize casualties, seems to be a vital component of any prudent security policy. What is at stake—maintaining national interests, promoting military credibility, fostering international stability, and protecting human life—is simply far too important to postpone any longer a thorough reconsideration of the quest for bloodless war.

Notes

1. Robert Mandel, "Fighting Fire with Fire: Privatizing Counterterrorism," in Russell D. Howard and Reid L. Sawyer, eds., *Defeating Terrorism: Shaping the New Security Environment* (New York: McGraw-Hill, 2003).
2. Harvey M. Sapolsky and Jeremy Shapiro, "Casualties, Technology, and America's Future Wars," *Parameters* (summer 1996): 121.
3. Jeffrey Record, "Force-Protection Fetishism: Sources, Consequences, and (?) Solutions," *Aerospace Power Journal* (summer 2000): 10.
4. Max Boot, *The Savage Wars of Peace: Small Wars and the Rise of American Power* (New York: Basic Books, 2002), pp. 347–348.
5. Admiral Gene LaRocque, "Approaching the Digital Battlefield" (Washington, DC: Center for Defense Information, December 15, 1996), available online at www.cdi.org/adm/transcripts/1014.
6. Sapolsky and Shapiro, "Casualties, Technology, and America's Future Wars," p. 124.

7. John N. Sims Jr., "Shackled by Perceptions: America's Desire for Bloodless Intervention," thesis presented to the faculty of the School of Advanced Airpower Studies, Maxwell Air Force Base, AL, June 1997, p. 79.

8. Colonel Charles J. Dunlap Jr., "Law and Military Interventions: Preserving Humanitarian Values in Twenty-First-Century Conflicts" (Washington, DC: Harvard University, John F. Kennedy School of Government, Carr Center for Human Rights Policy, Humanitarian Challenges in Military Intervention Conference, November 29, 2001), p. 2.

9. Peter D. Feaver and Christopher Gelpi, *Choosing Your Battles: American Civil-Military Relations and the Use of Force* (Princeton: Princeton University Press, 2004), pp. 206–207.

10. Karl W. Eikenberry, "Take No Casualties," *Parameters* (summer 1996): 115.

11. Andrew P. N. Erdmann, "The U.S. Presumption of Quick, Costless Wars," *Orbis* 43 (summer 1999): 379.

12. Jeffrey Record, *Failed States and Casualty Phobia: Implications for Force Structure and Technology Choices,* Occasional Paper no. 18 (Montgomery, AL: Center for Strategy and Technology, Air War College, September 2000), p. 23.

13. Sapolsky and Shapiro, "Casualties, Technology, and America's Future Wars," p. 121.

14. Boot, *Savage Wars of Peace,* pp. 347–348.

Selected Bibliography of Works on Bloodless War

Baylis, John, James Wirtz, Eliot Cohen, and Colin S. Gray, eds. *Strategy in the Contemporary World*. New York: Oxford University Press, 2002.

Boot, Max. *The Savage Wars of Peace: Small Wars and the Rise of American Power*. New York: Basic Books, 2002.

Burk, James. "Public Support for Peacekeeping in Lebanon and Somalia: Assessing the Casualties Hypothesis." *Political Science Quarterly* 114 (spring 1999): 53–78.

Byman, Daniel, and Matthew Waxman. *The Dynamics of Coercion: American Foreign Policy and the Limits of Military Might*. New York: Cambridge University Press, 2002.

Conversino, Mark J. "Sawdust Superpower: The Perceptions of U.S. Casualty Tolerance in the Post–Gulf War Era." *Strategic Review* 25 (winter 1997): 15–23.

Correll, John T. "Casualties." *Air Force Magazine* 86 (June 2003): 48–52.

Dupuy, Colonel Trevor N. *The Evolution of Weapons and Warfare*. Indianapolis, IN: Bobbs-Merrill, 1980.

Eikenberry, Karl W. "Take No Casualties." *Parameters* (summer 1996): 109–118.

Erdmann, Andrew P. N. "The U.S. Presumption of Quick, Costless Wars." *Orbis* 43 (summer 1999): 363–381.

Feaver, Peter D., and Christopher Gelpi. *Choosing Your Battles: American Civil-Military Relations and the Use of Force*. Princeton: Princeton University Press, 2004.

French, Shannon. *The Code of the Warrior: Exploring Warrior Values Past and Present*. Lanham, MD: Rowman & Littlefield, 2003.

Fuller, J. F. C. *Armament and History: The Influence of Armament on History from the Dawn of Classical Warfare to the End of the Second World War*. New York: Da Capo Press, 1998.

Gartner, Scott Sigmund, and Gary M. Segura. "War, Casualties, and Public Opinion." *Journal of Conflict Resolution* 42 (June 1998): 278–300.

Hyde, Major Charles K. "Casualty Aversion: Implications for Policy Makers and Senior Military Officers." *Aerospace Power Journal* 14 (summer 2000): 17–27.

Johnson, James Turner. *Morality and Contemporary Warfare*. New Haven: Yale University Press, 1999.

Kull, Steven, and I. M. Destler. *Misreading the Public: The Myth of a New Isolationism.* Washington, DC: Brookings Institution, 1999.

Merom, Gil. *How Democracies Lose Small Wars: State, Society, and the Failures of France in Algeria, Israel in Lebanon, and the United States in Vietnam.* New York: Cambridge University Press, 2003.

Mueller, Karl P. "Politics, Death, and Morality in U.S. Foreign Policy." *Aerospace Power Journal* 14 (summer 2000): 12–16.

Record, Jeffrey. "Collapsed Countries, Casualty Dread, and the New American Way of War." *Parameters* (summer 2002): 4–23.

Sapolsky, Harvey M., and Jeremy Shapiro. "Casualties, Technology, and America's Future Wars." *Parameters* (summer 1996): 119–127.

Index

Afghanistan war (Operation Enduring Freedom, 2001–2002), 17, 20, 29–30, 36, 38, 39, 59, 61, 62, 67, 70, 75, 77, 78, 82–83, 85, 111, 139, 184, 201
American way of war, 31, 73
Anarchy, international, 15–16, 47, 63, 65, 108, 191, 193
Automated warfare, 48, 53, 57–60, 65, 194, 198

Bloodless war: dangers, 155–169; definitional issues, 8–11; history, 26–39; means of pursuit, 45–65; motivations, 11–22
Brutality/brute force, 14, 28, 68, 70–72, 90, 91, 113, 161, 166, 176, 188

Casualty aversion/minimization/avoidance. *See* Bloodless war
Casualties, definition of, 9–16
Civilian-combatant distinction, 9, 16, 49, 75, 85, 89–90, 105, 108–109, 179, 183, 184
CNN effect, 18, 21

Collateral damage, 2, 9, 17, 19, 20, 21, 73, 74, 75, 89, 91, 103, 109, 157, 175
Command-and-control systems, 21, 53, 74, 88, 131, 145, 150, 173
Communication, 3, 18, 50–51, 63, 110, 119, 125, 126, 128–129, 131, 133, 136, 146, 148, 192–194

Democracy, 3, 12, 13–14, 15–16, 18, 55, 62, 75, 85, 118–119, 167, 190, 192, 193, 196, 198, 200, 201
Deterrence, 50–51, 52, 53, 65, 117, 167, 188
Disarmament, 46–47, 52, 53, 64–65

Economic sanctions, 47, 75, 106–107, 119

Force protection, 10, 16, 20, 24, 26, 27, 54, 55–56, 57, 106, 162–163, 185, 194

Geneva Conventions, 14–15
Globalization/interdependence, 16, 18, 125, 126, 150
Gulf War (Operation Desert Storm, 1991), 25–26, 33–34, 74, 78, 79–80, 85, 110, 138–139, 141, 162–163

Humanitarian issues, 11–16, 24, 47, 48, 61, 67, 75, 103, 104, 118–119, 177–178
Human life, protection of, 11, 12, 13, 16, 21, 27, 47, 51, 53, 54, 55, 64, 75, 86, 92, 103, 127, 134, 156, 159, 160, 167, 169, 178, 180–181, 192, 193, 194, 201
Human rights, 1, 15, 17, 63, 113, 114–115

Information disruption, 128–130, 132, 133–134, 136, 146–148
Information revolution, 18, 135–136
Information warfare: conditional utility, 146–150; dangers, 141–145; definition, 128–131; general, 50–51, 52, 53, 57, 64, 67, 99, 107, 113, 125–150, 155, 157, 161, 163, 167, 173, 174, 176, 180, 181, 183, 190, 193, 196, 197, 198, 199; history, 135–141; motives, 131–135
Intelligence, 49, 60, 85, 88–89, 92, 118, 178–179, 183, 184, 187–189
Irrationality, 23, 51, 52, 90–91, 120, 149, 176, 184
Iraq war (Operation Iraqi Freedom, 2003), 9, 15, 17, 35, 36, 37, 39, 61, 67, 70, 77, 78, 79, 83–85, 111–112, 139–141, 159, 163, 168, 184, 201

Just war theory, 12, 26

Kosovo war (Operation Allied Force, 1999), 10, 15, 38, 78, 80–81, 85, 139, 141, 162–163

Law, international. *See* Rules of war
Leadership, 8, 10, 13, 17, 22–23, 24, 26, 38, 54, 55, 87, 91, 165, 166, 168, 195–196, 200
Legitimacy, 1, 3, 12, 14, 49, 56, 63, 104, 174–175, 176, 189, 193
Lethality, 7, 19, 53, 76, 101, 104, 158, 160

Mechanized warfare. *See* Automated warfare
Media coverage, 1, 16, 18–19, 21, 55, 103, 104–105, 142, 148
Mercenaries. *See* Private military
Military warrior ethos, 3, 164–165, 192
Morality, 1, 3, 11–13, 15–16, 21–22, 46, 47, 52, 63, 133, 146, 166, 169, 189, 192, 198

Nonlethal weaponry: conditional utility, 116–121; dangers, 112–116; definition, 100–102; general, 49, 52, 53, 57, 64, 67, 99–121, 127, 132, 147, 148, 155, 161, 166, 167, 173, 174, 176, 180, 181, 183, 190, 193,

196, 198, 199; history, 107–112; motives, 102–107
Nonstate forces, 9, 16, 85, 90, 92, 105, 107, 108, 131, 135–136, 143, 175–176, 179, 184, 191
Norms, global security, 3, 4, 12–13, 15, 47, 85, 113, 191–192, 201

Offensive-defensive balance, 19, 46, 50, 77, 129–130, 144–145, 161–162
Outsourcing security. *See* Private military

Pentagon attack (2001). *See* September 11 terrorist attack
Postconflict reconstruction, 103, 168, 197–198
Precision-guided munitions: conditional utility, 88–92; dangers, 86–88; definition, 68–73; general, 19, 31, 49, 52, 53, 57, 64, 67–93, 99, 107, 110, 113, 127, 132, 147, 148, 155, 157, 160, 161, 163, 166, 167, 173, 174, 176, 180, 181, 183, 190, 193, 196, 197, 198, 199; history, 76–85; motivations, 73–75
Private military, 27, 48, 52–53, 60–63, 194, 198
Psychological operations, 31, 50, 127, 128, 129, 130, 132–133, 134, 136–137, 148–150, 183
Public opinion, 1, 9, 10, 13, 14, 16, 18, 22–26, 30, 55, 56, 62, 75, 103–104, 159, 167–168, 192, 199

Quest for bloodless war. *See* Bloodless war

Robotic warfare. *See* Automated warfare
Rules of war, 14–15, 16, 18, 46, 47–48, 52–53, 65, 113, 119, 178, 184, 194, 198

September 11 terrorist attack (2001), 8, 10, 12, 26, 36, 68, 99–100, 193
Smart bombs. *See* Precision-guided munitions
Somalia, U.S. intervention in, 23, 26, 32, 34, 37, 55, 107, 108, 110, 158, 177, 200

Targeting, 21, 68, 77, 88–89, 92, 126
Technology. *See* Weapons technology
Terrorism, global, 8, 9, 10, 12, 16, 17, 20, 26, 29–30, 36, 38, 39, 51, 59, 62, 68, 75, 90, 92, 99–100, 107, 108, 109, 111, 113–114, 125, 127, 129, 131, 139, 144, 175–176, 179, 182, 183, 184, 189, 193, 195

United Nations, 17, 110, 191

Vietnam War, 18, 29, 30, 31, 32–33, 37, 73, 74, 75, 77, 111, 157

War, bloodless. *See* Bloodless war
Weapons technology, 1, 3, 16, 19–21, 21–22, 29, 46, 47, 49,

56, 57, 76–78, 88, 92, 109–110, 163–164, 180–181, 196

World Trade Center bombing (2001). *See* September 11 terrorist attack

About the Book

In recent decades, government and military officials alike have pushed increasingly in the direction of "bloodless wars," where confrontations are undertaken—and ultimately won—with minimum loss of human life. Robert Mandel provides the first comprehensive analysis of this trend.

After exploring the moral, legal, military, and political bases of the desire to minimize wartime casualties, Mandel examines the actual strategies and tools involved; here the focus is on nonlethal weapons, precision-guided munitions, and information warfare. He then addresses the sobering practical constraints on aspirations to minimize casualties. His concluding review of policy options draws lessons from premodern patterns of warfare and calls for a more realistic understanding of the strategies available in today's security environment.

Robert Mandel is professor of international affairs at Lewis and Clark College. His numerous publications include *Armies Without States: The Privatization of Security* and *Deadly Transfers and the Global Playground: Transnational Security Threats in a Disorderly World*.